# Practice Problems wi

# A Guide for Learning Engineering Fluid Mechanics
## Ninth Edition

**Clayton T. Crowe**
*Washington State University-Pullman*

**Donald F. Elger**
*University of Idaho-Moscow*

John Wiley & Sons, Inc.

Cover Photo:   ©Bo Tornvig/AgeFotostock America, Inc.

Copyright © 2009, 2006 John Wiley & Sons, Inc. All rights reserved.

No part of this publication may be reproduced, stored in a retrieval system or transmitted in any form or by any means, electronic, mechanical, photocopying, recording, scanning, or otherwise, except as permitted under Sections 107 or 108 of the 1976 United States Copyright Act, without either the prior written permission of the Publisher, or authorization through payment of the appropriate per-copy fee to the Copyright Clearance Center, Inc., 222 Rosewood Drive, Danvers, MA 01923, or on the web at www.copyright.com. Requests to the Publisher for permission should be addressed to the Permissions Department, John Wiley & Sons, Inc., 111 River Street, Hoboken, NJ 07030-5774, (201) 748-6011, fax (201) 748-6008, or online at http://www.wiley.com/go/permissions.

To order books or for customer service, please call 1-800-CALL-WILEY (225-5945).

ISBN-13   978- 0470-42086-7

Printed in the United States of America

10 9 8 7 6 5 4 3 2 1

Printed and bound by Integrated Book Technologies.

# Contents

1 Introduction — 1
2 Fluid Properties — 5
3 Fluid Statics — 9
4 Flowing Fluids and Pressure Variation — 19
5 Control Volume Approach and Continuity Equation — 29
6 Momentum Principle — 41
7 Energy Principle — 55
8 Dimensional Analysis and Similitude — 65
9 Surface Resistance — 75
10 Flow in Conduits — 87
11 Drag and Lift — 99
12 Compressible Flow — 109
13 Flow Measurements — 117
14 Turbomachinery — 125
15 Varied Flow in Open Channels — 131

# Preface

This volume presents a variety of example problems for students of fluid mechanics. It is a companion manual to the text, *Engineering Fluid Mechancis, 9$^{th}$ Edition* by Clayton T. Crowe, Donald F. Elger, Barbara C. Williams, and John A. Roberson.

Andrew DuBuisson, Steven Ruzich, and Ashley Ater have provided help with editing and with checking the solutions for accuracy.

Please transmit any comments or recommendations to

Professor Donald F. Elger
Mechanical Engineering Department
University of Idaho
Moscow, ID 83844-0902
delger@uidaho.edu
tel: 208-885-7889
fax: 208-885-9031

# Chapter 1

# Introduction

## Problem 1.1

Consider a glass container, half-full of water and half-full of air, at rest on a laboratory table. List some similarities and differences between the liquid (water) and the gas (air).

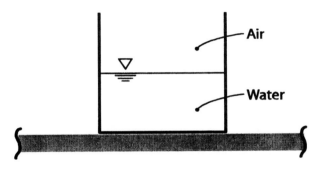

### Solution

Similarities

1. The gas and the liquid are comprised of molecules.

2. The gas and the liquid are fluids.

3. The molecules in the gas and the liquid are relatively free to move about.

4. The molecules in each fluid are in continual and random motion.

Differences

1. In the liquid phase, there are strong attractive and repulsive forces between the molecules; in the gas phase (assuming ideal gas), there are minimal forces between molecules except when they are in close proximity (mutual repulsive forces simulate collisions).

2. A liquid has a definite volume; a gas will expand to fill its container. Since the container is open in this case, the gas will continually exchange molecules with the ambient air.

3. A liquid is much more viscous than a gas.

4. A liquid forms a free surface, whereas a gas does not.

5. Liquids are very difficult to compress (requiring large pressures for small compression), whereas gases are relatively easy to compress.

6. With the exception of evaporation, the liquid molecules stay in the container. The gas molecules constantly pass in and out of the container.

7. A liquid exhibits an evaporation phenomenon, whereas a gas does not.

## Comments

Most of the differences between gases and liquids can be understood by considering the differences in molecular structure. Gas molecules are far apart, and each molecule moves independently of its neighbor, except when one molecule approaches another. Liquid molecules are close together, and each molecule exerts strong attractive and repulsive forces on its neighbor.

# Problem 1.2

In an ink-jet printer, the orifice that is used to form ink drops can have a diameter as small as $3 \times 10^{-6}$ m. Assuming that ink has the properties of water, does the continuum assumption apply?

## Solution

The continuum assumption will apply if the size of a volume, which contains enough molecules so that effects due to random molecular variations average out, is much smaller that the system dimensions. Assume that $10^4$ molecules is sufficient for averaging. If $L$ is the length of one side of a cube that contains $10^4$ molecules and $D$ is the diameter of the orifice, the continuum assumption is satisfied if

$$\frac{L}{D} \ll 1$$

The number of molecules in a mole of matter is Avogadro's number: $6.02 \times 10^{23}$. The molecular weight of water is 18, so the number of molecules ($N$) in a gram of water is

$$N = \left(\frac{6.02 \times 10^{23} \text{ molecules}}{\text{mole}}\right)\left(\frac{\text{mole}}{18 \text{ g}}\right)$$
$$= 3.34 \times 10^{12} \frac{\text{molecules}}{\text{g}}$$

The density of water is 1 g/cm³, so the number of molecules in a cm³ is $3.34 \times 10^{12}$. The volume of water that contains $10^4$ molecules is

$$\text{Volume} = \frac{10^4 \text{ molecules}}{3.34 \times 10^{12} \frac{\text{molecules}}{\text{cm}^3}}$$
$$= 3.0 \times 10^{-19} \text{ cm}^3$$

Since the volume of a cube is $L^3$, where $L$ is the length of a side

$$L = \sqrt[3]{3.0 \times 10^{-19} \text{ cm}^3}$$
$$= 6.2 \times 10^{-7} \text{ cm}$$
$$= 6.2 \times 10^{-9} \text{ m}$$

Thus

$$\frac{L}{D} = \frac{6.2 \times 10^{-9} \text{ m}}{3.0 \times 10^{-6} \text{ m}}$$
$$= 0.0021$$

Since $\frac{L}{D} \ll 1$, *the continuum assumption is quite good.*

# Chapter 2

# Fluid Properties

## Problem 2.1

Calculate the density and specific weight of nitrogen at an absolute pressure of 1 MPa and a temperature of 40°C.

## Solution

Ideal gas law

$$\rho = \frac{p}{RT}$$

From Table A.2, $R = 297$ J/kg/K. The temperature in absolute units is $T = 273 + 40 = 313$ K.

$$\rho = \frac{10^6 \text{ N/m}^2}{297 \text{ J/kgK} \times 313 \text{ K}}$$
$$= 10.75 \text{ kg/m}^3$$

The specific weight is

$$\gamma = \rho g$$
$$= 10.76 \text{ kg/m}^3 \times 9.81 \text{ m/s}^2$$
$$= 105.4 \text{ N/m}^3$$

# Problem 2.2

Find the density, kinematic and dynamic viscosity of crude oil in traditional units at 100°F.

### Solution

From Fig. A.3, $\nu = 6.5 \times 10^{-5}$ ft$^2$/s and $S = 0.86$.

The density of water at standard condition is 1.94 slugs/ft$^3$, so the density of crude oil is $0.86 \times 1.94 = 1.67$ slugs/ft$^3$ or $1.67 \times 32.2 = 53.8$ lbm/ft$^3$.

The dynamic viscosity is $\rho\nu = 1.67 \times 6.5 \times 10^{-5} = 1.09 \times 10^{-4}$ lbf·s/ft$^2$.

# Problem 2.3

Two parallel glass plates separated by 0.5 mm are placed in water at 20°C. The plates are clean, and the width/separation ratio is large so that end effects are negligible. How far will the water rise between the plates?

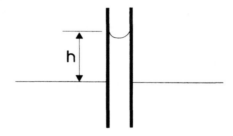

### Solution

The surface tension at 20°C is $7.3 \times 10^{-2}$ N/m. The weight of the water in the column $h$ is balanced by the surface tension force.

$$whd\rho g = 2w\sigma \cos\theta$$

where $w$ is the width of the plates and $d$ is the separation distance. For water against glass, $\cos\theta \simeq 1$. Solving for $h$ gives

$$h = \frac{\sigma}{d\rho g} = \frac{2 \times 7.3 \times 10^{-2} \text{ N/m}}{0.5 \times 10^{-3} \text{ m} \times 998 \text{ kg/m}^3 \times 9.81 \text{ m/s}^2}$$
$$= 0.0149 \text{ m} = \underline{29.8 \text{ mm}}$$

# Problem 2.4

The kinematic viscosity of helium at 15°C and standard atmospheric pressure (101 kPa) is $1.14\times10^{-4}$ m$^2$/s. Using Sutherland's equation, find the kinematic viscosity at 100°C and 200 kPa.

## Solution

From Table A.2, Sutherland's constant for helium is 79.4 K and the gas constant is 2077 J/kgK. Sutherland's equation for absolute viscosity is

$$\frac{\mu}{\mu_o} = \left(\frac{T}{T_o}\right)^{3/2} \frac{T_o + S}{T + S}$$

The absolute viscosity is related to the kinematic viscosity by $\mu = \nu\rho$. Substituting into Sutherland's equation

$$\frac{\rho\nu}{\rho_o\nu_o} = \left(\frac{T}{T_o}\right)^{3/2} \frac{T_o + S}{T + S}$$

or

$$\frac{\nu}{\nu_o} = \frac{\rho_o}{\rho}\left(\frac{T}{T_o}\right)^{3/2} \frac{T_o + S}{T + S}$$

From the ideal gas law

$$\frac{\rho_o}{\rho} = \frac{p_o}{p}\frac{T}{T_o}$$

so

$$\frac{\nu}{\nu_o} = \frac{p_o}{p}\left(\frac{T}{T_o}\right)^{5/2} \frac{T_o + S}{T + S}$$

The kinematic viscosity ratio is found to be

$$\frac{\nu}{\nu_o} = \frac{101}{200}\left(\frac{373}{288}\right)^{5/2} \frac{288 + 79.4}{373 + 79.4}$$
$$= 0.783$$

The kinematic viscosity is

$$\nu = 0.783 \times 1.14 \times 10^{-4} = \underline{\underline{8.93 \times 10^{-5} \text{ m}^2/\text{s}}}$$

# Problem 2.5

Air at 15°C forms a boundary layer near a solid wall. The velocity distribution in the boundary layer is given by

$$\frac{u}{U} = 1 - \exp(-2\frac{y}{\delta})$$

where $U = 30$ m/s and $\delta = 1$ cm. Find the shear stress at the wall ($y = 0$).

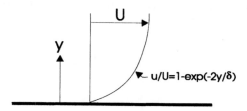

## Solution

The shear stress at the wall is related to the velocity gradient by

$$\tau = \mu \frac{du}{dy}\Big|_{y=0}$$

Taking the derivative with respect to $y$ of the velocity distribution

$$\frac{du}{dy} = 2\frac{U}{\delta}\exp(-2\frac{y}{\delta})$$

Evaluating at $y = 0$

$$\frac{du}{dy}\Big|_{y=0} = 2\frac{U}{\delta} = 2 \times \frac{30}{0.01} = 6 \times 10^3 \text{ s}^{-1}$$

From Table A.2, the density of air is 1.22 kg/m³, and the kinematic viscosity is $1.46 \times 10^{-5}$ m²/s. The absolute viscosity is $\mu = \rho\nu = 1.22 \times 1.46 \times 10^{-5} = 1.78 \times 10^{-5}$ N·s/m². The shear stress at the wall is

$$\tau = \mu\frac{du}{dy}\Big|_{y=0} = 1.78 \times 10^{-5} \times 6 \times 10^3 = \underline{\underline{0.107 \text{ N/m}^2}}$$

# Chapter 3

# Fluid Statics

## Problem 3.1

For a lake, find the depth $h$ at which the gage pressure is 1 atmosphere. The specific weight of water is 62.3 lbf/ft$^3$.

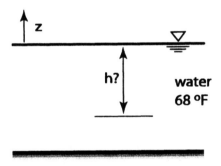

### Solution

At the free surface of the lake, pressure will be $p_{\text{surface}} = 1.0$ atm absolute or 0.0 atm gage. At a depth $h$, the pressure will be 1 atm gage.

In a static fluid of constant density, the piezometric head $(p/\gamma + z)$ is constant. Thus

$$\frac{p_{\text{surface}}}{\gamma} + z_{\text{surface}} = \frac{1 \text{ atm}}{\gamma} + (z_{\text{surface}} - h) \tag{1}$$

Since $p_{surface} = 0$ atm gage, Eq. (1) becomes

$$h = \frac{1 \text{ atm}}{\gamma}$$

$$= \frac{(14.7 \text{ lbf/in}^2)(144 \text{ in}^2/\text{ft}^2)}{62.3 \text{ lbf/ft}^3}$$

$$= \underline{34.0 \text{ ft}}$$

## Problem 3.2

A tank that is open to the atmosphere contains a 1.0-m layer of oil ($\rho = 800$ kg/m$^3$) floating on a 0.5-m layer of water ($\rho = 1000$ kg/m$^3$). Determine the pressure at elevations $A$, $B$, $C$, and $D$. Note that $B$ is midway between $A$ and $C$.

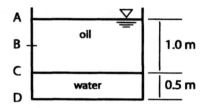

## Solution

At a horizontal interface of two fluids, pressure will be constant across the interface. Thus the pressure in the oil at $A$ equals the pressure in the air (atmospheric pressure).

$$p_A = p_{atm}$$
$$= \underline{0 \text{ kPa gage}}$$

Since the oil layer is a static fluid of constant density, the piezometric pressure is constant

$$p_A + \gamma_{oil} z_A = p_B + \gamma_{oil} z_B = p_C + \gamma_{oil} z_C = \text{constant} \qquad (1)$$

where $z$ denotes elevation. Let $z_A = 0$, $z_B = -0.5$ m, $z_C = -1.0$ m. Then, Eq. (1) becomes

$$p_A = p_B + \gamma_{oil}(-0.5 \text{ m}) = p_C + \gamma_{oil}(-1.0 \text{ m})$$

So

$$p_B = p_A + \gamma_{oil}(0.5 \text{ m})$$
$$= p_{atm} + (800)(9.81)(0.5)/1000$$
$$= \underline{3.92 \text{ kPa-gage}}$$

Similarly
$$p_C = p_A + \gamma_{oil}(1.0 \text{ m})$$
$$= p_{atm} + (800)(9.81)(1.0)/1000$$
$$= \underline{\underline{7.85 \text{ kPa-gage}}}$$

At elevation C, pressure in the oil equals pressure in the water. Since the piezometric pressure in the water is constant, we can write

$$p_C + \gamma_{water} z_C = p_D + \gamma_{water} z_D$$

or

$$p_D = p_C + \gamma_{water}(z_C - z_D)$$
$$= 7.85 + (1000)(9.81)(0.5)/1000$$
$$= \underline{\underline{12.8 \text{ kPa-gage}}}$$

## Problem 3.3

A U-tube manometer contains kerosene, mercury and water, each at 70 °F. The manometer is connected between two pipes ($A$ and $B$), and the pressure difference, as measured between the pipe centerlines, is $p_B - p_A = 4.5$ psi. Find the elevation difference $z$ in the manometer.

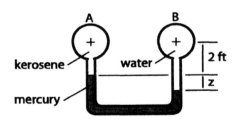

### Solution

Apply the manometer equation (Eq. 3.17 in the 8th edition). Begin at location $B$ and add pressure differences until location $A$ is reached.

$$p_B + (2+z)\gamma_{water} - z\gamma_{Hg} - 2\gamma_{kero} = p_A$$

Rearranging
$$p_B - p_A = 2(\gamma_{kero} - \gamma_{water}) + z(\gamma_{Hg} - \gamma_{water}) \qquad (1)$$

Looking up values of specific weight and substituting into Eq. (1) gives

$$[4.5 \times 144] \text{ lbf/ft}^2 = \left[(2 \text{ ft})(51 - 62.3) \text{ lbf/ft}^3 + (z \text{ ft})(847 - 62.3) \text{lbf/ft}^3\right]$$

So
$$z = \frac{(648 + 22.6)}{784.7}$$
$$= \underline{\underline{0.855 \text{ ft}}}$$

---

# Problem 3.4

A container, filled with water at 20°C, is open to the atmosphere on the right side. Find the pressure of the air in the enclosed space on the left side of the container.

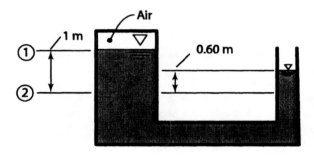

---

## Solution

The pressure at elevation 2 is the same on both the left and right side.

$$p_2 = p_{\text{atm}} + \gamma (0.6 \text{ m})$$
$$= 0 + (9.81 \text{ kN/m}^3)(0.6 \text{ m})$$
$$= 5.89 \text{ kPa}$$

Since the piezometric head is the same at elevations 1 and 2

$$\frac{p_1}{\gamma} + z_1 = \frac{p_2}{\gamma} + z_2$$

so

$$p_1 = p_2 + \gamma (z_2 - z_1)$$
$$= (5.89 \text{ kPa}) + (9.81 \text{ kN/m}^3)(-1.0 \text{ m})$$
$$= \underline{\underline{-3.92 \text{ kPa gage}}}$$

## Problem 3.5

A rectangular gate of dimension 1 m by 4 m is held in place by a stop block at B. This block exerts a horizontal force of 40 kN and a vertical force of 0 kN. The gate is pin-connected at $A$, and the weight of the gate is 2 kN. Find the depth $h$ of the water.

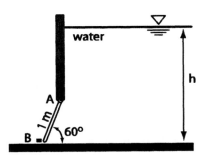

## Solution

A free-body diagram of the gate is

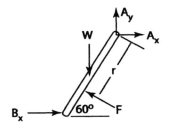

where $W$ is the weight of the gate, $F$ is the equivalent force of the water, and $r$ is the length of the moment arm. Summing moments about $A$ gives

$$B_x(1.0\sin 60°) - F \times r + W(0.5\cos 60°) = 0$$

or

$$\begin{aligned} F \times r &= B_x \sin 60° + W(0.5\cos 60°) \\ &= 40,000 \sin 60° + 2000(0.5\cos 60°) \\ &= 35,140 \text{ N-m} \end{aligned} \quad (1)$$

The hydrostatic force $F$ acts at a distance $\bar{I}/\bar{y}A$ below the centroid of the plate. Thus the length of the moment arm is

$$r = 0.5 \text{ m} + \frac{\bar{I}}{\bar{y}A} \quad (2)$$

Analysis of terms in Eq. (2) gives

$$\bar{y} = (h/\sin(60°) - 0.5)$$
$$\bar{I} = 4 \times 1^3/12 = 0.333$$
$$A = 4 \times 1 = 4$$

Eq. (2) becomes

$$r = 0.5 + \frac{0.0833}{(h/\sin(60°) - 0.5)} \quad (3)$$

The equivalent force of the water is

$$F = \bar{p}A$$
$$= \gamma(h - 0.5\sin 60°)4$$
$$= 9,810(h - 0.5\sin 60°)4$$
$$= 39,240(h - 0.433) \quad (4)$$

Substituting Eqs. (3) and (4) into Eq. (1) gives

$$35,140 = Fr$$
$$35,140 = [39,240(h - 0.433)]\left[0.5 + \frac{0.0833}{(1.155h - 0.5)}\right] \quad (5)$$

Eq. (5) has a single unknown (the depth of water $h$). To solve Eq. (5), one may use a computer program that finds the root of an equation. This was done, and the answer is

$$h = \underline{2.08 \text{ m}}$$

## Problem 3.6

A container is formed by joining two plates, each 4 ft long with a dimension of 6 ft in the direction normal to the paper. The plates are joined by a pin connection at $A$ and held together at the top by two steels rods (one on each end). The container is filled with concrete ($S = 2.4$) to a depth of 1.5 ft. Find the tensile load in each steel rod.

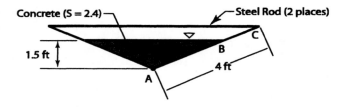

## Solution

A free-body diagram of plate $ABC$ is

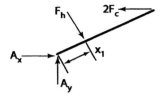

Summing moments about point $A$

$$F_h x_1 = 2F_c \left(4 \sin(30°)\text{ ft}\right)$$

or

$$F_c = \frac{F_h x_1}{4 \text{ ft}} \tag{1}$$

The length from $A$ to $B$ is $\overline{AB} = 1.5/\cos(60°) = 3$ ft. The hydrostatic force $(F_h)$ is the product of area $AB$ and pressure of the concrete at a depth of 0.75 ft.

$$\begin{aligned}F_h &= \left(\overline{AB} \times 6 \text{ ft}\right)(\gamma_{\text{concrete}})(0.75 \text{ ft}) \\ &= \left(3 \times 6 \text{ ft}^2\right)\left(2.4 \times 62.4 \text{ lbf/ft}^3\right)(0.75 \text{ ft}) \\ &= 2020 \text{ lbf}\end{aligned}$$

The geometry of plate $ABC$ is

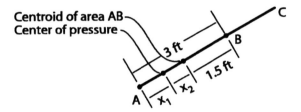

To find the distance $x_2$, note that portion $BC$ of the plate is above the surface of the concrete. Thus use values for a plate of dimension 3 ft by 6 ft.

$$\begin{aligned}x_2 &= \frac{\overline{I}}{\overline{y}A} \\ &= \frac{\left(6 \text{ ft} \times 3^3 \text{ ft}^3\right)/12}{(1.5 \text{ ft})\left(3 \times 6 \text{ ft}^2\right)} \\ &= 0.5 \text{ ft}\end{aligned}$$

The moment arm $x_1$ is

$$\begin{aligned}x_1 &= (1.5 \text{ ft}) - x_2 \\ &= 1.0 \text{ ft}\end{aligned}$$

Eq. (1) becomes

$$F_c = \frac{F_h x_1}{4 \text{ ft}}$$
$$= \frac{(2020 \text{ lbf})(1.0 \text{ ft})}{4 \text{ ft}}$$
$$= \underline{505 \text{ lbf}}$$

## Problem 3.7

A closed glass tube (hydrometer) of length $L$ and diameter $D$ floats in a reservoir filled with a liquid of unknown specific gravity $S$. The glass tube is partially filled with air and partially filled with a liquid that has a specific gravity of 3. Determine the specific gravity of the reservoir fluid. Neglect the weight of the glass walls of the tube.

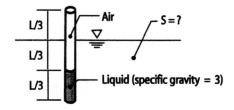

### Solution

A free-body diagram is

The buoyant force on the tube is

$$F_B = S\gamma_{H_2O} V_{\text{Displaced}} = S\gamma_{H_2O} \left( \frac{\pi D^2}{4} \times \frac{2L}{3} \right)$$

Weight of the fluid in the tube is

$$W = 3\gamma_{H_2O} V_{\text{Liquid}} = 3\gamma_{H_2O}\left(\frac{\pi D^2}{4} \times \frac{L}{3}\right)$$

From the equilibrium principle, weight balances the buoyant force.

$$W = F_B$$

$$3\gamma_{H_2O}\left(\frac{\pi D^2}{4} \times \frac{L}{3}\right) = S\gamma_{H_2O}\left(\frac{\pi D^2}{4} \times \frac{2L}{3}\right)$$

Eliminating common terms

$$3 = S \times 2$$

Thus

$$\underline{\underline{S = 1.5}}$$

## Problem 3.8

An 18-in. diameter concrete cylinder ($S = 2.4$) is used to raise a 60-ft long log to a 45° angle. The center of the log is pin-connected to a pier at point $A$. Find the length $L$ of the concrete cylinder.

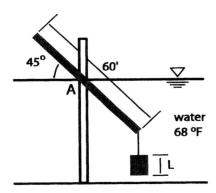

## Solution

A free-body diagram is

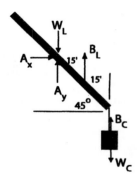

where $B_L$ and $B_C$ are the buoyant forces on the log and concrete, respectively. Similarly, $W_L$ and $W_C$ represent weight.

Summing moments about point $A$

$$B_L\,(15')\cos 45° + (B_C - W_C)\,(30')\cos 45° = 0 \tag{1}$$

The buoyant force on the log is

$$\begin{aligned} B_L &= \gamma_{H_2O} V_{Disp} = \gamma_{H_2O}\left(\frac{\pi D_L^2}{4}\times 30'\right) \\ &= 62.3\left(\frac{\pi 1^2}{4}\times 30'\right) \\ &= 1470\ \text{lbf} \end{aligned} \tag{2}$$

The net force on the concrete is

$$\begin{aligned} F_{net} &= B_C - W_C \\ &= \gamma_{H_2O}V_{Concrete} - \gamma_{Concrete}V_{Concrete} \\ &= \gamma_{H_2O}(1 - S_{concrete})V_{Concrete} \\ &= 62.3(1-2.4)\left(\frac{\pi 1.5^2}{4}\times L\right) \\ &= -154.1L\ \text{lbf} \end{aligned} \tag{3}$$

Combining Eqs. (1) to (3)

$$(1467\ \text{lbf})(15\ \text{ft})\cos 45° - (154.1L\ \text{lbf})(30\ \text{ft})\cos 45° = 0$$

Thus

$$\underline{L = 4.76\ \text{ft}}$$

# Chapter 4

# Flowing Fluids and Pressure Variation

## Problem 4.1

A flow moves in the $x$-direction with a velocity of 10 m/s from 0 to 0.1 second and then reverses direction with the same speed from 0.1 to 0.2 second. Sketch the pathline starting from $x = 0$ and the streakline with dye introduced at $x = 0$. Show the streamlines for the first time interval and the second time interval.

## Solution

The pathline is the line traced out by a fluid particle released from the origin. The fluid particle first goes to $x = 1.0$ and then returns to the origin so the pathline is

The streakline is the configuration of the dye at the end of 0.2 second. During the first period, the dye forms a streak extending from the origin to $x = 1$ m. During the second period, the whole field moves to the left while dye continues to be injected. The final configuration is a line extending from the origin to $x = -1$ m.

The streamlines are represented by

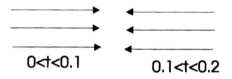

## Problem 4.2

A piston is accelerating upward at a rate of 10 m/s². A 50-cm-long water column is above the piston. Determine the pressure at a distance of 20 cm below the water surface. Neglect viscous effects.

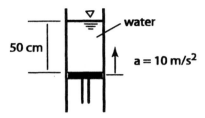

## Solution

Since the water column is accelerating, Euler's equation applies. Let the $\ell$-direction be coincident with elevation, that is, the $z$-direction. Euler's equation becomes

$$-\frac{\partial}{\partial z}(p + \gamma z) = \rho a_z \qquad (1)$$

Since pressure varies with $z$ only, the left side of Euler's equation becomes

$$-\frac{\partial}{\partial z}(p + \gamma z) = -\left(\frac{dp}{dz} + \gamma\right) \qquad (2)$$

Combining Eqs. (1) and (2) gives

$$\begin{aligned}\frac{dp}{dz} &= -(\rho a_z + \gamma) \\ &= -(1000 \text{ kg/m}^3 \times 10 \text{ m/s}^2 + 9810 \text{ N/m}^3) \\ &= -19,810 \text{ N/m}^3\end{aligned} \qquad (3)$$

Integrating Eq. (3) from the water surface ($z = 0$ m) to a depth of 20 cm ($z = -0.2$ m) gives

$$\int_{p(z=0)}^{p(z=-0.2)} dp = \int_{z=0}^{z=-0.2} -19.8 \; dz$$

$$p(z = -0.2) - p(z = 0) = \left(-19.8 \text{ kN/m}^3\right)(-0.2 - 0) \text{ m}$$

Since pressure at the water surface ($z = 0$) is 0,

$$p_{z=-0.2 \text{ m}} = \left(-19.8 \text{ kN/m}^3\right)(-0.2 \text{ m})$$

$$= \underline{\underline{3.96 \text{ kPa-gage}}}$$

# Problem 4.3

A rectangular tank, initially at rest, is filled with kerosene ($\rho = 1.58$ slug/ft$^3$) to a depth of 4 ft. The space above the kerosene contains air that is at a pressure of 0.8 atm. Later, the tank is set in motion with a constant acceleration of 1.2 g to the right. Determine the maximum pressure in the tank after the onset of motion.

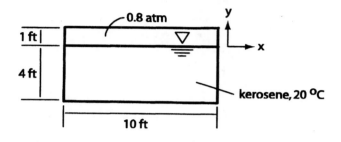

## Solution

After initial sloshing is damped out, the configuration of the kerosene is shown in Fig. 1.

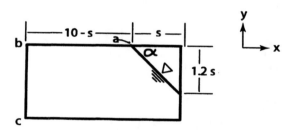

Figure 1 Configuration of kerosene during acceleration

In Fig. 1, $s$ is an unknown length, and the angle $\alpha$ is

$$\tan \alpha = \frac{a_x}{g} = 1.2$$

So

$$\alpha = 50.2°$$

To find the length $s$, note that the volume of the air space before and after motion remains constant.

$$(10)(1)(\text{width}) = (1/2)(1.2s)(s)(\text{width})$$

So

$$s = \sqrt{20/1.2} = 4.08 \text{ ft}$$

The maximum pressure will occur at point $c$ in Fig. 1. Before finding this pressure, find the pressure at $b$ by integrating Euler's equation from point $a$ to point $b$.

$$-\frac{dp}{dx} = \rho a_x$$

$$\int_a^b dp = -\int_a^b \rho a_x dx$$

$$p_b = p_a + \rho a_x (10 - s)$$
$$= (0.8)(2116.2) \text{ lbf/ft}^2 + (1.58)(1.2 \times 32.2)(10 - 4.08) \text{ lbf/ft}^2$$
$$= 2054 \text{ lbf/ft}^2 \text{ absolute}$$

To find the pressure at $c$, Euler's equation may be integrated from $b$ to $c$.

$$-\frac{d(p + \gamma z)}{dz} = 0$$
$$\frac{dp}{dz} = -\gamma$$
$$\int_b^c dp = \int_b^c -\gamma dz$$
$$p_c - p_b = -\gamma(z_c - z_b)$$

so

$$p_c = p_b + \gamma(5 \text{ ft})$$
$$= 2054 \text{ lbf/ft}^2 + (1.58 \times 32.2 \text{ lbf/ft}^3)(5 \text{ ft})$$
$$= \underline{\underline{2310 \text{ lbf/ft}^2\text{-absolute}}}$$

# Problem 4.4

A cylindrical tank contains air at a density of 1.2 kg/m³. The pressure in the tank is maintained at constant value such that the air exiting the 2-cm diameter nozzle has a constant speed of 25 m/s. Determine the pressure value as indicated by the Bourdon-tube gage at the top of the tank. Assume irrotational flow.

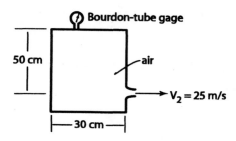

## Solution

When flow is irrotational, the Bernoulli equation applies. Apply this equation between location 1 at the gage and location 2 on the exit plane of the jet.

$$\frac{p_1}{\gamma} + \frac{V_1^2}{2g} + z_1 = \frac{p_2}{\gamma} + \frac{V_2^2}{2g} + z_2 \qquad (1)$$

At location the air would be barely moving so $V_1 \approx 0$. Define a datum at the elevation of the nozzle; thus $z_1 = 0.5$ m, and $z_2 = 0$. Pressure across a subsonic air jet is atmospheric; thus $p_2 = 0$ gage. Eq. (1) becomes

$$\frac{p_1}{\gamma} = \frac{V_2^2}{2g} - z_1$$

or

$$\begin{aligned} p_1 &= \frac{\rho V_2^2}{2} - \rho g z_1 \\ &= \frac{(1.2)(25)^2}{2} - (1.2)(9.81)(0.5) \\ &= \underline{\underline{369 \text{ Pa-gage}}} \end{aligned}$$

Note that the elevation terms are quite small. When applying the Bernoulli equation to a gas, elevation terms are commonly neglected.

# Problem 4.5

An airfoil is being tested in an open channel flow of water at 60°F. The velocity at point $A$ is twice the approach velocity $V$. Determine the maximum value of the approach velocity such that cavitation does not occur.

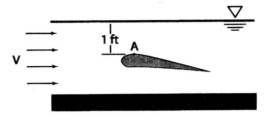

## Solution

Cavitation will occur when the pressure at $A$ equals the vapor pressure of water at 60°F. From Table A.5

$$p_A = 0.256 \text{ psia}$$
$$= 36.9 \text{ psfa}$$

Identify locations 1 to 3 as shown by the points in the following sketch.

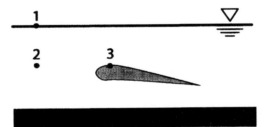

The Bernoulli equation between 2 and 3 is

$$\frac{p_2}{\gamma} + \frac{V^2}{2g} = \frac{36.9 \text{ psfa}}{\gamma} + \frac{(2V)^2}{2g} \qquad (1)$$

Apply the Bernoulli equation between 1 and 2. Since $V_1 = V_2$, and $p_1 = p_\text{atm}$, the Bernoulli equation simplifies to the hydrostatic condition.

$$\frac{p_\text{atm}}{\gamma} + (1 \text{ ft}) = \frac{p_2}{\gamma}$$

Substituting values gives

$$p_2 = p_\text{atm} + \gamma (1 \text{ ft})$$
$$= (2116 \text{ psf}) + \left(62.37 \text{ lbf/ft}^3\right)(1 \text{ ft}) \qquad (2)$$
$$= 2180 \text{ psfa}$$

Combining Eqs. (1) and (2) gives

$$(2180 \text{ psfa}) + \frac{\rho V^2}{2} = (36.9 \text{ psfa}) + \frac{\rho(2V)^2}{2}$$

So

$$2140 = \frac{1.94 V^2 (2^2 - 1)}{2}$$
$$2140 = 2.91 V^2$$

or

$$V = \underline{\underline{27.1 \text{ ft/s}}}$$

## Problem 4.6

A u-tube filled with mercury ($\rho = 13{,}550$ kg/m$^3$) is rotated about axis *A-A*. Length *L* is 25 cm and the column height *z* is 5 cm. Determine the rotation speed ($\omega$).

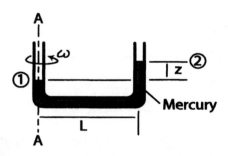

### Solution

Integration of Euler's (see Eq. 5.9 in the 7$^{\text{th}}$ edition) shows that $p + \gamma z - \rho r^2 \omega^2/2 =$ constant. Thus

$$\left(p + \gamma z - \rho r^2 \omega^2/2\right)_1 = \left(p + \gamma z - \rho r^2 \omega^2/2\right)_2$$

where locations 1 and 2 denote the liquid surfaces. Locate an elevation datum along surface 1. Then

$$p_1 = p_2 + \gamma z - \rho L^2 \omega^2/2$$

Since $p_1 = p_2 = 0$ kPa-gage,

$$\gamma z = \rho L^2 \omega^2 / 2$$

or

$$\omega = \sqrt{\frac{2gz}{L^2}}$$
$$= \sqrt{\frac{2 \times 9.81 \times 0.05}{0.25^2}}$$
$$= \underline{\underline{3.96 \text{ rad/s}}}$$

# Chapter 5

# Control Volume Approach and Continuity Principle

## Problem 5.1

A 10-cm-diameter pipe contains sea water that flows with a mean velocity of 5 m/s. Find the volume flow rate (discharge) and the mass flow rate.

### Solution

The discharge is

$$Q = VA$$

where $V$ is the mean velocity. Thus

$$Q = 5 \times \frac{\pi}{4} \times 0.1^2$$
$$= 0.0393 \text{ m}^3/\text{s}$$

From Table A.4, the density of sea water is 1026 kg/m$^3$.

The mass flow rate is

$$\dot{m} = \rho Q = 1026 \times 0.0393 = \underline{\underline{40.3 \text{ kg/s}}}$$

# Problem 5.2

The velocity profile of a non-Newtonian fluid in a circular conduit is given by

$$\frac{u}{u_{max}} = \left[1 - \left(\frac{r}{R}\right)^2\right]^{1/2}$$

where $u_{max}$ is the velocity at the centerline and $R$ is the radius of the conduit. Find the discharge (volume flow rate) in terms of $u_{max}$ and $R$.

# Solution

The volume flow rate is

$$Q = \int_A u \, dA$$

For an axisymmetric duct, this integral can be written as

$$Q = 2\pi \int_0^R u r \, dr$$

Substituting in the equation for the velocity distribution

$$Q = 2\pi u_{max} \int_0^R \left[1 - \left(\frac{r}{R}\right)^2\right]^{1/2} r \, dr$$

Recognizing that $2r\,dr = dr^2$, we can rewrite the integral as

$$Q = \pi u_{max} \int_0^R \left[1 - \left(\frac{r}{R}\right)^2\right]^{1/2} dr^2$$

$$= \pi u_{max} R^2 \int_0^R \left[1 - \left(\frac{r}{R}\right)^2\right]^{1/2} d\left(\frac{r}{R}\right)^2$$

or

$$Q = \pi u_{max} R^2 \int_0^1 [1 - \eta]^{1/2} d\eta$$

$$= -\frac{2}{3} \pi u_{max} R^2 [1 - \eta]^{3/2} \Big|_0^1$$

$$= \underline{\underline{\frac{2}{3} \pi u_{max} R^2}}$$

# Problem 5.3

A jet pump injects water at 120 ft/s through a 2-in. pipe into a secondary flow in an 8-in. pipe where the velocity is 10 ft/s. Downstream the flows become fully mixed with a uniform velocity profile. What is the magnitude of the velocity where the flows are fully mixed?

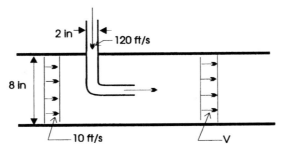

## Solution

Draw a control volume as shown in the sketch below.

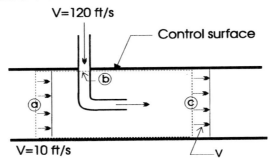

Because the flow is steady

$$\sum_{cs} \rho \mathbf{V} \cdot \mathbf{A} = 0$$

Assuming the water is incompressible, the continuity equation becomes

$$\sum_{cs} \mathbf{V} \cdot \mathbf{A} = 0$$

The volume flow rate across station $a$ is

$$\sum_a \mathbf{V} \cdot \mathbf{A} = -10 \times \frac{\pi}{4} \left(\frac{8}{12}\right)^2$$

where the minus sign occurs because the velocity and area vectors have the opposite sense. The volume flow rate across station $b$ is

$$\sum_b \mathbf{V} \cdot \mathbf{A} = -120 \times \frac{\pi}{4} \left(\frac{2}{12}\right)^2$$

and the volume flow rate across station $c$ is

$$\sum_c \mathbf{V} \cdot \mathbf{A} = V \times \frac{\pi}{4}\left(\frac{8}{12}\right)^2$$

where $V$ is the velocity. Substituting into the continuity equation

$$-120 \times \frac{\pi}{4}\left(\frac{2}{12}\right)^2 - 10 \times \frac{\pi}{4}\left(\frac{8}{12}\right)^2 + V \times \frac{\pi}{4}\left(\frac{8}{12}\right)^2 = 0$$

$$V = \frac{(120 \times 2^2 + 10 \times 8^2)}{8^2}$$

$$= \underline{\underline{17.5 \text{ ft/s}}}$$

## Problem 5.4

Water flows into a cylindrical tank at the rate of 1 m$^3$/min and out at the rate of 1.2 m$^3$/min. The cross-sectional area of the tank is 2 m$^2$. Find the rate at which the water level in the tank changes. The tank is open to the atmosphere.

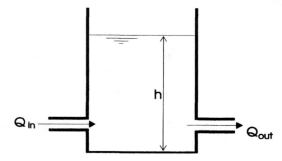

### Solution

Draw a control volume around the fluid in the tank. Assume the control surface moves with the free surface of the water.

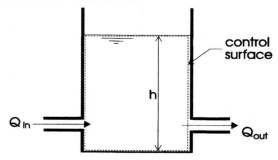

The continuity equation is

$$\frac{d}{dt}\int_{cv}\rho d\forall + \sum_{cs}\rho \mathbf{V}\cdot\mathbf{A} = 0$$

The density inside the control volume is constant so

$$\frac{d}{dt}\int_{cv} d\forall + \sum_{cs}\mathbf{V}\cdot\mathbf{A} = 0$$

$$\frac{d\forall}{dt} + \sum_{cs}\mathbf{V}\cdot\mathbf{A} = 0$$

The volume of the fluid in the tank is $\forall = hA$. Mass crosses the control surface at two locations. At the inlet

$$\mathbf{V}\cdot\mathbf{A} = -Q_{in}$$

and at the outlet

$$\mathbf{V}\cdot\mathbf{A} = Q_{out}$$

Substituting into the continuity equation

$$A\frac{dh}{dt} + Q_{out} - Q_{in} = 0$$

or

$$\begin{aligned}\frac{dh}{dt} &= \frac{Q_{in} - Q_{out}}{A} \\ &= \frac{1 - 1.2}{2} \\ &= \underline{\underline{-0.1 \text{ m/min}}}\end{aligned}$$

# Problem 5.5

Water flows steadily through a nozzle. The nozzle diameter at the inlet is 2 in., and the diameter at the exit is 1.5 in. The average velocity at the inlet is 5 ft/s. What is the average velocity at the exit?

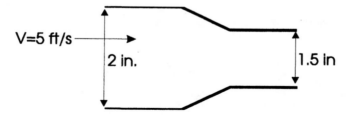

## Solution

Because the flow is steady, the continuity equation is

$$\sum_{cs} \rho \mathbf{V} \cdot \mathbf{A} = 0$$

Also, because the fluid is incompressible, the continuity equation reduces to

$$\sum_{cs} \mathbf{V} \cdot \mathbf{A} = 0$$

Draw a control surface that includes the inlet and outlet sections of the nozzle as shown.

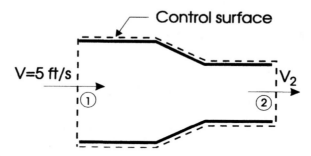

At the inlet, station 1,

$$(\mathbf{V} \cdot \mathbf{A})_1 = -5 \times \frac{\pi}{4} \times \left(\frac{2}{12}\right)^2$$

At the exit, station 2,

$$(\mathbf{V} \cdot \mathbf{A})_2 = V_2 \times \frac{\pi}{4} \times \left(\frac{1.5}{12}\right)^2$$

Substituting into the continuity equation

$$\sum_{cs} \mathbf{V} \cdot \mathbf{A} = -5 \times \frac{\pi}{4} \times \left(\frac{2}{12}\right)^2 + V_2 \times \frac{\pi}{4} \times \left(\frac{1.5}{12}\right)^2 = 0$$

or

$$V_2 = 5 \times \left(\frac{2.0}{1.5}\right)^2 = \underline{\underline{8.89 \text{ ft/s}}}$$

## Problem 5.6

Air flows steadily through a 10 cm-diameter conduit. The velocity, pressure, and temperature of the air at station 1 are 30 m/s, 100 kPa absolute, and 300 K. At station 2, the pressure has decreased to 95 kPa absolute, and the temperature remains constant between the two stations (isothermal flow). Find the mass flow rate and the velocity at station 2.

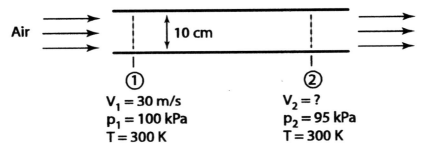

### Solution

The mass flow rate is

$$\dot{m} = \rho V A$$

The density is obtained from the equation of state for an ideal gas.

$$\rho = \frac{p}{RT}$$

At station 1

$$\rho_1 = \frac{100 \times 10^3 \text{ N/m}^2}{287 \text{ J/kgK} \times 300 \text{ K}} = 1.16 \text{ kg/m}^3$$

The flow rate is

$$\dot{m} = 1.16 \times 30 \times \frac{\pi}{4} \times 0.1^2 = \underline{\underline{0.273 \text{ kg/s}}}$$

Because the flow is steady, the continuity equation reduces to

$$\sum_{cs} \rho \mathbf{V} \cdot \mathbf{A} = 0$$

# 36 CHAPTER 5. CONTROL VOLUME APPROACH AND CONTINUITY PRINCIPLE

which states that the rate of mass flow through station 1 will be equal to that through station 2. The air density at station 2 is

$$\rho_2 = \frac{95 \times 10^3 \text{ N/m}^2}{287 \text{ J/kgK} \times 300 \text{ K}} = 1.10 \text{ kg/m}^3$$

The mass flow is the same at each station. Thus

$$(\rho V A)_1 = (\rho V A)_2$$

So

$$V_2 = V_1 \frac{\rho_1}{\rho_2} = 30 \times \frac{1.16}{1.10} = \underline{\underline{31.6 \text{ m/s}}}$$

## Problem 5.7

Water flows steadily through a 4-cm diameter pipe that is 10-m long. The pipe wall is porous, leading to a small flow through the pipe wall. The inlet velocity is 10 m/s, and the exit velocity is 9 m/s. Find the average velocity of the water that is passing through the porous surface.

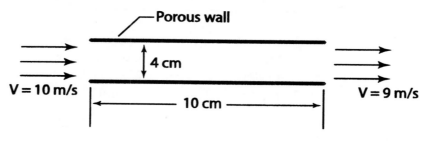

## Solution

The flow rate is steady, and the fluid is incompressible so the continuity equation reduces to

$$\sum_{cs} \mathbf{V} \cdot \mathbf{A} = 0$$

Draw a control surface around the pipe. The entrance is station 1, the exit is station 2, and the surface of the porous pipe is station 3.

For station 1

$$(\mathbf{V} \cdot \mathbf{A})_1 = -10 \times \frac{\pi}{4} \times 0.04^2$$

For station 2
$$(\mathbf{V} \cdot \mathbf{A})_2 = 9 \times \frac{\pi}{4} \times 0.04^2$$

For the porous surface
$$(\mathbf{V} \cdot \mathbf{A})_3 = V_3 \times \pi \times 0.04 \times 10$$

The continuity equation is
$$-10 \times \frac{\pi}{4} \times 0.04^2 + 9 \times \frac{\pi}{4} \times 0.04^2 + V_3 \times \pi \times 0.04 \times 10 = 0$$

or
$$V_3 = \underline{\underline{0.001 \text{ m/s}}}$$

## Problem 5.8

Water is forced out of a 2-cm diameter nozzle by a 6-cm-diameter piston moving at a speed of 5 m/s. Determine the force required to move the piston and the speed of the fluid jet ($V_2$). Neglect friction on the piston and assume irrotational flow. The exit pressure ($p_2$) is atmospheric.

### Solution

When flow is irrotational, the Bernoulli equation applies. Applying this equation along the nozzle centerline between locations 1 and 2 gives
$$\frac{p_1}{\gamma} + \frac{V_1^2}{2g} = \frac{p_2}{\gamma} + \frac{V_2^2}{2g} \tag{1}$$

The continuity principle is
$$A_1 V_1 = A_2 V_2$$

So
$$V_2 = (5 \text{ m/s}) \frac{0.06^2}{0.02^2}$$
$$= 45 \text{ m/s} \tag{2}$$

Letting $p_2 = 0$ kPa gage and combining Eqs. (1) and (2)
$$p_1 = \rho \frac{V_2^2 - V_1^2}{2}$$
$$= (1000) \frac{45^2 - 5^2}{2}$$
$$= 1 \text{ MPa}$$

Since the piston is moving at a constant speed, the applied force $F$ is balanced by the pressure force.

$$F = p_1 A_1$$
$$= (1 \text{ MPa}) \left(\pi \times 0.06^2/4 \text{ m}^2\right)$$
$$= \underline{2.83 \text{ kN}}$$

## Problem 5.9

The sketch shows a fertilizer sprayer that uses a Venturi nozzle. Water moving through this nozzle reaches a low pressure at section 1. This low pressure draws liquid fertilizer (assume fertilizer has the properties of water) up the suction tube, and the mixture is jetted to ambient at section 2. Nozzle dimensions are $d_1 = 3$ mm, $d_2 = 9$ mm, and $h = 150$ mm. Determine the minimum possible water speed ($V_2$) at the exit of the nozzle so that fluid will be drawn up the suction tube.

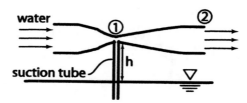

## Solution

Since we are looking for the lower limit of operation, assume inviscid flow so that the Bernoulli equation applies. Also assume that the pressure at 1 is just low enough to draw fluid up the suction tube, meaning there is no flow in the suction tube.

Identify locations 1 to 3 as shown by the points in the sketch below.

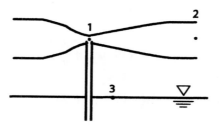

Applying the hydrostatic principle (constant piezometric pressure in a fluid of constant density) between 1 and 3 gives

$$p_3 = p_1 + \gamma(z_1 - z_3)$$

Let $p_3 = 0$ kPa gage, and let $(z_1 - z_3) = (h + d_1/2)$.

$$p_1 = -\gamma(h + d_1/2) \tag{1}$$

Applying the Bernoulli equation between 1 and 2 gives

$$\frac{p_1}{\gamma} + \frac{V_1^2}{2g} = \frac{p_2}{\gamma} + \frac{V_2^2}{2g} \tag{2}$$

The continuity principle is

$$A_1 V_1 = A_2 V_2 \tag{3}$$

Let $p_2 = 0$ kPa gage, and combine Eqs. (2) and (3).

$$p_1 = \frac{\rho V_2^2}{2}\left(1 - \frac{d_2^4}{d_1^4}\right) \tag{4}$$

Combine Eqs. (1) and (4).

$$-\gamma(h + d_1/2) = \frac{\rho V_2^2}{2}\left(1 - \frac{d_2^4}{d_1^4}\right)$$

$$-9800(0.15 + 0.003/2) = \frac{1000 \times V_2^2}{2}\left(1 - \left(\frac{9}{3}\right)^4\right)$$

So

$$V_2 = \underline{\underline{0.193 \text{ m/s}}}$$

# Problem 5.10

Show that the velocity field

$$\mathbf{V} = xyz^2\mathbf{i} - \frac{y^2}{2}z\mathbf{j} + y\left(\frac{z^2}{2} - \frac{z^3}{3}\right)\mathbf{k}$$

satisfies the continuity equation for an incompressible flow and find the vorticity at the point (1,1,1).

## Solution

The continuity equation for the flow of an incompressible fluid is

$$\nabla \cdot \mathbf{V} = \frac{\partial u}{\partial x} + \frac{\partial v}{\partial y} + \frac{\partial w}{\partial z} = 0$$

Substituting in the velocity derivatives

$$yz^2 - yz + yz - yz^2 \equiv 0$$

so continuity equation is satisfied.

The equation for vorticity is

$$\omega = \nabla \times \mathbf{V}$$

$$= \mathbf{i}\left(\frac{\partial w}{\partial y} - \frac{\partial v}{\partial z}\right) + \mathbf{j}\left(\frac{\partial u}{\partial z} - \frac{\partial w}{\partial x}\right) + \mathbf{k}\left(\frac{\partial v}{\partial x} - \frac{\partial u}{\partial y}\right)$$

Substituting in the velocity derivatives

$$\omega = \mathbf{i}\left[\left(\frac{z^2}{2} - \frac{z^3}{3}\right) + \frac{y^2}{2}\right] + \mathbf{j}\left(2xy - 0\right) + \mathbf{k}\left(0 - xz^2\right)$$

Substituting values at point (1,1,1)

$$\underline{\underline{\omega = \frac{2}{3}\mathbf{i} + 2\mathbf{j} - \mathbf{k}}}$$

# Chapter 6

# Momentum Principle

## Problem 6.1

Water at 20°C is discharged from a nozzle onto a plate as shown. The flow rate of the water is 0.001 m³/s, and the diameter of the nozzle outlet is 0.5 cm. Find the force necessary to hold the plate in place.

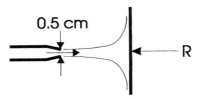

### Solution

This is a one-dimensional, steady flow. Since the system is not accelerating, the velocities with respect to the nozzle and plate are inertial velocities. The momentum equation in the $x$-direction (horizontal direction) is

$$\sum F_x = \sum \dot{m}_o v_{o_x} - \sum \dot{m}_i v_{i_x}$$

Draw a control volume with the associated force and momentum diagrams.

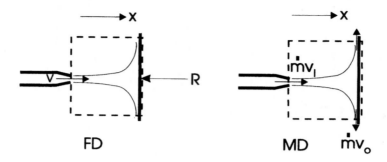

From the force diagram

$$\sum F_x = -R$$

From the continuity equation, the mass flow in is equal to the mass flow out so

$$\dot{m}_o = \dot{m}_i = \dot{m}$$

The velocity at the inlet is $V$. The component of velocity in the $x$-direction at the outlet is zero, so the momentum flux is

$$\sum \dot{m}_o v_{o_x} - \sum \dot{m}_i v_{i_x} = -\dot{m}V$$

Equating the forces and momentum flux

$$-R = -\dot{m}V$$

or

$$R = \dot{m}V$$

The volume flow rate is 0.01 m³/s, so the mass flow rate is $\dot{m} = \rho Q = 1000 \times 0.001 = 1$ kg/s. The velocity is

$$V = \frac{Q}{A} = \frac{0.001}{\frac{\pi}{4}(0.005)^2} = 50.9 \text{ m/s}$$

The restraining force is

$$R = 50.9 \times 1 = \underline{50.9 \text{ N}}$$

# Problem 6.2

A water jet with a velocity of 30 m/s impacts on a splitter plate so that $\frac{1}{4}$ of the water is deflected toward the bottom and $\frac{3}{4}$ toward the top. The angle of the plate is 45°. Find the force required to hold the plate stationary. Neglect the weight of the plate and water, and neglect viscous effects.

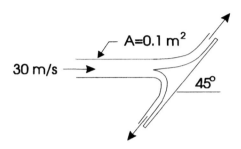

# Solution

The pressure is constant on the free surface of the water. Because frictional effects are neglected, the Bernoulli equation is applicable. Without gravitational effects, the Bernoulli equation becomes

$$p + \frac{1}{2}\rho V^2 = \text{constant}$$

Since pressure is constant, the velocity will be constant. Therefore, each exit velocity is equal to the inlet velocity.

Momentum and force diagrams for this problem are

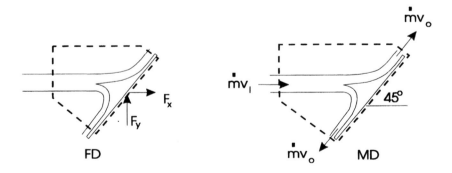

The forces acting on the control surface are

$$\mathbf{F} = F_x \mathbf{i} + F_y \mathbf{j}$$

The momentum flux from the momentum diagram is

$$\sum_{cs} \dot{m}_o \mathbf{v}_o - \sum_{cs} \dot{m}_i \mathbf{v}_i = \frac{3}{4}\dot{m}_i(30\mathbf{i}\cos 45 + 30\mathbf{j}\sin 45)$$
$$+ \frac{1}{4}\dot{m}_i(-30\mathbf{i}\cos 45 - 30\mathbf{j}\sin 45) - \dot{m}_i 30\mathbf{i}$$

Equating the force and momentum flux

$$F_x\mathbf{i} + F_y\mathbf{j} = \dot{m}_i(-19.4\mathbf{i}+10.6\mathbf{j})$$

The inlet mass flow rate is

$$\dot{m}_i = \rho A V = 1000 \times 0.1 \times 30 = 3000 \text{ kg/s}$$

The force vector evaluates to

$$F_x\mathbf{i} + F_y\mathbf{j} = -5.82 \times 10^4 \mathbf{i} + 3.18 \times 10^4 \mathbf{j} \text{ (N)}$$

Thus

$$F_x = \underline{\underline{-5.82 \times 10^4 \text{ N}}}$$
$$F_y = \underline{\underline{3.18 \times 10^4 \text{ N}}}$$

## Problem 6.3

A 12-in. horizontal pipe is connected to a reducer to a 6-in. pipe. Crude oil flows through the pipe at a rate of 10 cfs. The pressure at the inlet to the reducer is 60 psi. Find the force of the fluid on the reducer. The specific gravity of crude oil is 0.86. The Bernoulli equation can be used through the reducer.

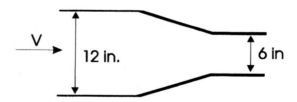

## Solution

Draw a force and momentum diagram as shown.

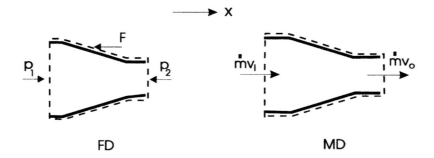

FD  MD

At the control surface, forces due to pressure are acting. Also, a force ($F$) is needed to hold the reducer stationary.

$$\sum_{cs} F_x = p_1 A_1 - p_2 A_2 - F \qquad (1)$$

The control volume is not accelerating and the flow is steady, so the momentum change becomes

$$\sum_{cs} \dot{m}_o v_{o,x} - \sum_{cs} \dot{m}_i v_{i,x} = \dot{m} V_2 - \dot{m} V_1 = \dot{m}(V_2 - V_1) \qquad (2)$$

The velocity at the inlet is

$$V_1 = \frac{Q}{A_1} = \frac{10}{\frac{\pi}{4} 1^2} = 12.7 \text{ ft/s}$$

The velocity at the exit is

$$V_2 = \frac{Q}{A_2} = \frac{10}{\frac{\pi}{4}(0.5)^2} = 50.9 \text{ ft/s}$$

The pressure at the outlet can be found by applying the Bernoulli equation.

$$\frac{p_1}{\gamma} + \frac{V_1^2}{2g} = \frac{p_2}{\gamma} + \frac{V_2^2}{2g}$$

So

$$p_2 = p_1 + \frac{\rho}{2}(V_1^2 - V_2^2)$$
$$= 60 \times 144 + \frac{0.86 \times 1.94}{2}(12.7^2 - 50.9^2)$$
$$= 8640 - 2027 = 6613 \text{ psf} = 45.9 \text{ psi}$$

Equating the force and momentum flux by combining Eqs. (1) and (2) gives

$$60 \times \frac{\pi}{4} \times 12^2 - 45.9 \times \frac{\pi}{4} \times 6^2 - F = 10 \times 0.86 \times 1.94 \times (50.9 - 12.7)$$
$$F = \underline{4850 \text{ lbf}}$$

# Problem 6.4

An eductor is a pump with no moving parts in which a high-speed jet is injected into a slower moving fluid. In the eductor shown in the following figure, water is injected through a 2-cm nozzle at a speed of 30 m/s. The flow of water in the 5-cm duct is 5 m/s. If the pressure downstream where the flow is totally mixed is 100 kPa, what is the pressure where the water is injected through the nozzle? Neglect the friction on the walls.

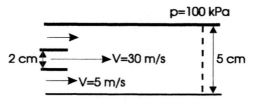

## Solution

Draw a control surface as shown with the appropriate force and momentum diagrams.

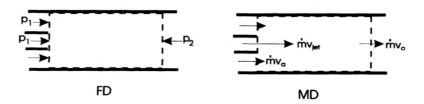

The velocity at the outlet may be obtained from the continuity equation for a steady flow. The mass flow in the high-speed jet is

$$\dot{m}_{jet} = 1000 \times \frac{\pi}{4} \times 30 \times 0.02^2$$
$$= 9.42 \text{ kg/s}$$

The mass flow through the outer annular region is

$$\dot{m}_a = 1000 \times \frac{\pi}{4} \times 5 \times (0.05^2 - 0.02^2)$$
$$= 8.24 \text{ kg/s}$$

The velocity at the outlet is

$$V_o = (\dot{m}_{jet} + \dot{m}_a)/\rho A$$
$$= (9.42 + 8.24)/(1000 \times \frac{\pi}{4} \times 0.05^2)$$
$$= 9 \text{ m/s}$$

The sum of the forces on the control surface is

$$\sum_{cs} F_x = p_1 A - p_2 A$$
$$= \frac{\pi}{4} \times 0.05^2 \times (p_1 - 10^5)$$
$$= 0.00196 \times (p_1 - 10^5) \text{ N}$$

The control surface is not accelerating and the flow is steady, so the momentum flux is

$$\sum_{cs} \dot{m}_o v_{o,x} - \sum_{cs} \dot{m}_i v_{i,x} = (9.42 + 8.24) \times 9 - 9.42 \times 30 - 8.24 \times 5$$
$$= -164.86 \text{ kg} \cdot \text{m/s}^2$$

Equating the force and momentum flux

$$0.00196 \times (p_1 - 10^5) = -164.86$$
$$= \underline{15.9 \text{ kPa}}$$

## Problem 6.5

A turbojet with a 1-m diameter inlet is being tested in a facility capable of simulating high-altitude conditions where the atmospheric pressure is 55 kPa absolute and the temperature is 267 K. The gas constant for air is 287 J/kg/K. The velocity at the inlet is 100 m/s. The exit diameter is 0.75 m, the exit temperature is 800 K, and the exit pressure is the local atmospheric pressure. Find the thrust produced by the turbojet.

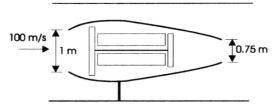

### Solution

Draw the force and momentum diagrams as shown.

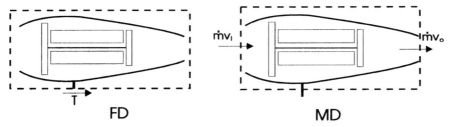

From the force diagram

$$\sum F_x = T$$

where $T$ is the thrust, which is the force applied to the strut in the free-body diagram.

From the momentum diagram

$$\sum \dot{m}_o v_{o,x} - \sum \dot{m}_i v_{i,x} = \dot{m}(v_{o,x} - v_{i,x})$$

since the flow is steady and the mass flow in equals the mass flow out. Equating the force and momentum gives

$$T = \dot{m}(v_{o,x} - v_{i,x})$$

Since the control volume is stationary, the fluid velocities relative to the control volume are relative to an inertial reference frame; so

$$T = \dot{m}(V_o - V_i)$$

The density of the air at the inlet is

$$\rho_i = \frac{p_i}{RT_i} = \frac{55 \times 10^3}{287 \times 267} = 0.718 \text{ kg/m}^3$$

The mass flow is

$$\dot{m} = \rho A V = 0.718 \times \frac{\pi}{4} \times 1^2 \times 100 = 56.4 \text{ kg/s}$$

The density at the exit is

$$\rho_o = \frac{p_o}{RT_o} = \frac{55 \times 10^3}{287 \times 800} = 0.240 \text{ kg/m}^3$$

The outlet velocity is obtained from

$$V_o = \frac{\dot{m}}{\rho_o A_o} = \frac{56.4}{0.240 \times \frac{\pi}{4} \times 0.75^2} = 532 \text{ m/s}$$

The thrust is

$$T = 56.4 \times (532 - 100) = 24,360 \text{ N} = \underline{24.4 \text{ kN}}$$

# Problem 6.6

A retro-rocket is used to decelerate a rocket ship in space. The rocket is moving at 8000 m/s (with respect to the Earth's surface) and has a mass of 1000 kg. The burn rate of the retrorocket is 8 kg/s, and the exhaust velocity with respect to the rocket nozzle is 2000 m/s. After the retrorocket has fired, the velocity should be 7500 m/s. How long must the retro-rocket be fired, and what is the final mass of the rocket? Assume the exit pressure of the rocket is equal to the ambient pressure and the drag forces on the rocket are negligible. The rocket is moving in a direction perpendicular to the gravity force.

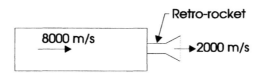

## Solution

Draw a control volume around the rocket as shown.

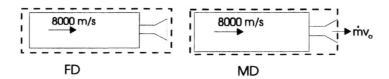

Because the exit pressure of the retrorocket is equal to the ambient pressure, there are no forces acting on the control surface. Also there is no body force in the direction of interest. From the force diagram

$$\sum F_x = 0$$

From the momentum diagram

$$\frac{d}{dt}\int_{cv} \rho v d\forall + \sum_{cs} \dot{m}_o v_o - \sum_{cs} \dot{m}_i v_i = \frac{d}{dt}\int_{cv} \rho v_x d\forall + \dot{m}_o v_{o,x}$$

where the velocities must be referenced to an inertial coordinate system. Equating the forces and momentum change gives

$$\frac{d}{dt}\int_{cv} \rho v_x d\forall + \dot{m}_o v_{o,x} = 0$$

The unsteady term can be written as

$$\frac{d}{dt}\int_{cv} \rho v_x d\forall = \frac{d}{dt}(v_R \int_{cv} \rho d\forall) = \frac{d}{dt}(m_R v_R)$$

where $v_R$ and $m_R$ are the velocity and mass of the rocket, respectively. The momentum flux term becomes

$$\dot{m}_o v_{o,x} = \dot{m}(v_R + V_e)$$

since the velocity must be referenced with respect to an inertial reference frame.

Finally, the equation becomes

$$\frac{d}{dt}(m_R v_R) + \dot{m}(v_R + V_e) = 0$$

$$m_R \frac{dv_R}{dt} + v_R \frac{dm_R}{dt} + \dot{m} v_R + \dot{m} V_e = 0$$

However, from the continuity equation $dm_R/dt = -\dot{m}$. Thus the equation reduces to

$$m_R \frac{dv_R}{dt} + \dot{m} V_e = 0 \tag{1}$$

The mass of the rocket will decrease linearly with time as

$$m_R = m_o - \dot{m} t$$

where $m_o$ is the initial mass of the rocket. Eq. (1) can now be rewritten as

$$dv_R = -V_e \frac{\dot{m} dt}{m_o - \dot{m} t}$$

Integrating

$$\Delta v_R = V_e \ln\left(1 - \frac{\dot{m} t}{m_o}\right)$$

Substituting in values

$$-500 = 2000 \times \ln(1 - \frac{8 \times t}{1000})$$
$$0.789 = 1 - 0.008 \times t$$
$$t = \underline{26.4 \text{ s}}$$

## Problem 6.7

A rotating arm with a radius of 1 meter has slits through which water issues with a uniform velocity of 10 m/s in the outer halves of the arms. The slits are 3 mm wide and 0.5 m long. The arm rotates with a constant angular velocity of 5 rad/s. Find the torque required to keep the arm rotating at this speed.

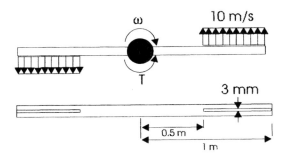

## Solution

This problem use the moment-of-momentum principle. The force and momentum diagrams are

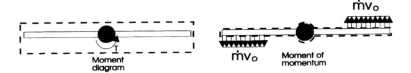

The only moment acting on the arm is the torque, which is a vector coming out of the page.

$$\sum \mathbf{M}_o = T\mathbf{k}$$

The flow is steady, so the moment of momentum is

$$\int_{cs} (\mathbf{r} \times \mathbf{v}) \rho V \, dA = \mathbf{k} \int_{cs} r v \rho V \, dA$$

because the velocity is perpendicular to the radius. Thus the moment of momentum equation becomes

$$T = \int_{cs} r v \rho V \, dA$$

The differential area is $w\,dr$. The velocity with respect to an inertial reference frame is

$$v = V - \omega r$$

Thus the equation for torque becomes

$$T = 2\rho w \int_{0.5}^{1} rV(V - \omega r)dr$$

$$= 2\rho w \left[\frac{1}{2}r^2 V^2 - \frac{1}{3}\omega V r^3\right]_{0.5}^{1}$$

$$= 2 \times 1000 \times 0.003 \times \left[\frac{10^2}{2} \times 0.75 - \frac{5 \times 10}{3} \times 0.875\right]$$

$$= \underline{137.5 \text{ N} \cdot \text{m}}$$

# Problem 6.8

A six-in. pipe is used to carry water for a distance of one mile (5280 ft). Before a valve is closed, the initial pressure in the pipe is 20 psig. Determine the maximum flow rate (in gpm) in the pipe so that when the valve is closed, the water hammer pressure will not exceed 50 psig. Also determine the critical closure time. The modulus of elasticity of water is 320,000 psi. The density of the water is 1.94 slugs/ft$^3$.

## Solution

The pressure increase in a pipe due to the water hammer effect is

$$\Delta p = \rho V c$$

where $c$ is the speed of sound in water. The speed of sound is calculated from

$$c = \sqrt{\frac{E_v}{\rho}}$$

$$= \sqrt{\frac{320,000 \times 144}{1.94}}$$

$$= 4874 \text{ ft/s}$$

The final pressure in the pipe is

$$p_f = p_i + \rho V c$$

where $p_i$ is the initial pressure. Therefore

$$\rho V c = p_f - p_i$$
$$= 500 - 20 = 480 \text{ psi} = 69{,}120 \text{ psf}$$

Solving for $V$

$$V = \frac{69{,}120}{1.94 \times 4874} = 7.31 \text{ ft/s}$$

The flow rate is

$$Q = AV = \frac{\pi}{4}\left(\frac{6}{12}\right)^2 \times 7.31 = 1.44 \text{ cfs} = \underline{\underline{646 \text{ gpm}}}$$

The critical closure time is

$$\begin{aligned} t_c &= \frac{2L}{c} \\ &= \frac{2 \times 5280}{4874} = \underline{\underline{2.17 \text{ s}}} \end{aligned}$$

If the time to close the valve is longer than 2.17 seconds, the pressure rise will be less.

# Chapter 7

# Energy Principle

## Problem 7.1

Air flows through a rectangular duct of dimension 1 ft × 5 ft. The velocity profile is linear, with a maximum velocity of 15 ft/s. Find the kinetic energy correction factor.

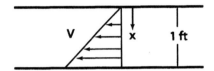

## Solution

The kinetic energy correction factor is given by

$$\alpha = \frac{1}{A} \int_A \frac{V^3}{\overline{V}^3} dA \qquad (1)$$

Area is $1 \times 5 = 5$ ft². Since the profile is linear, the average velocity is half the maximum velocity.

$$\overline{V} = V_{\max}/2$$
$$= 7.5 \text{ ft/s}$$

Pick a differential area with a height of $dx$ and width of 5 ft.

$$dA = 5dx$$

Eq. (1) becomes

$$\alpha = \frac{1}{5}\left(\frac{1}{7.5}\right)^3 \int_A V^3 \,(5dx)$$

$$= 0.00237 \int_{x=0}^{x=1 \text{ ft}} [V(x)]^3 \, dx \qquad (2)$$

Since $V(x)$ is a straight line, it can be fit with an equation of the form $V = mx + b$, where $m$ is slope and $b$ is intercept. The result is

$$V(x) = 15x \qquad (3)$$

Combining Eqs. (2) and (3)

$$\alpha = 0.00237\,(15^3) \int_{x=0}^{x=1 \text{ ft}} x^3 \, dx$$

$$= 0.00237\,(15^3)\,(1/4)$$

$$\alpha = \underline{\underline{2}}$$

# Problem 7.2

Water flows out of a large tank through a 1-cm diameter siphon tube. The siphon is terminated with a nozzle of diameter 3 mm. Determine the minimum pressure in the siphon and determine the velocity of the water leaving the siphon. Assume laminar flow and assume all energy losses due to effects of viscosity are negligible.

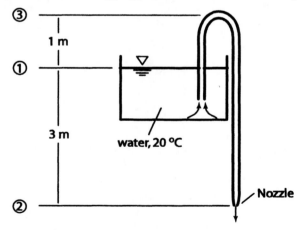

## Solution

Let location 1 be coincident with the free surface of the water in the tank. Let location 2 be the exit of the siphon. The energy equation between 1 and 2 is

$$\frac{p_1}{\gamma} + \alpha_1 \frac{\overline{V}_1^2}{2g} + z_1 + h_p = \frac{p_2}{\gamma} + \alpha_2 \frac{\overline{V}_2^2}{2g} + z_2 + h_t + h_L$$

Now $p_1 = p_2 = 0, \overline{V}_1 \approx 0$, $z_1 = 3$ m, $h_p = h_t = h_L = 0$. Since the flow is laminar, $\alpha_2 = 2$. The energy equation simplifies to

$$z_1 = \alpha_2 \frac{\overline{V}_2^2}{2g}$$

$$3 = 2\frac{\overline{V}_2^2}{2 \times 9.8}$$

So

$$\overline{V}_2 = \underline{\underline{5.42 \text{ m/s}}}$$

The minimum pressure will occur at the highest point in the siphon; let this be location 3. The energy equation between 1 and 3 is

$$\frac{p_1}{\gamma} + \alpha_1 \frac{\overline{V}_1^2}{2g} + z_1 + h_p = \frac{p_3}{\gamma} + \alpha_3 \frac{\overline{V}_3^2}{2g} + z_3 + h_t + h_L$$

Dropping terms that are zero and simplifying gives

$$z_1 = \frac{p_3}{\gamma} + \alpha_3 \frac{\overline{V}_3^2}{2g} + z_3 \tag{1}$$

To find $\overline{V}_3$, use the continuity principle.

$$\overline{V}_3 A_3 = \overline{V}_2 A_2$$

$$\overline{V}_3 = \overline{V}_2 \frac{A_2}{A_3}$$

$$= 5.42 \frac{0.003^2}{0.01^2}$$

$$= 0.488 \text{ m/s}$$

Substitute numbers into Eq. (1).

$$z_1 = \frac{p_3}{\gamma} + \alpha_3 \frac{\overline{V}_3^2}{2g} + z_3$$

$$3 = \frac{p_3}{9800} + 2\frac{0.488^2}{2 \times 9.81} + 4$$

$$-1.024 = \frac{p_3}{9800}$$

So

$$p_3 = \underline{\underline{-10.0 \text{ kPa}}}$$

## Problem 7.3

A pump with an efficiency of 70% pumps water at 60°F in a four-in. pipe. Determine the power required by the pump. Neglect head loss, and assume all kinetic energy correction factors are unity.

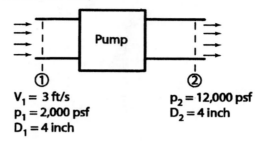

$V_1 = 3$ ft/s
$p_1 = 2,000$ psf
$D_1 = 4$ inch

$p_2 = 12,000$ psf
$D_2 = 4$ inch

## Solution

The energy equation between sections 1 and 2 is

$$\frac{p_1}{\gamma} + \alpha_1 \frac{\overline{V}_1^2}{2g} + z_1 + h_p = \frac{p_2}{\gamma} + \alpha_2 \frac{\overline{V}_2^2}{2g} + z_2 + h_t + h_L$$

By continuity, $\overline{V}_1 = \overline{V}_2$, and so the velocity head terms cancel. Also, $z_1 = z_2$. The energy equation simplifies to

$$\frac{p_1}{\gamma} + h_p = \frac{p_2}{\gamma}$$

or

$$h_p = \frac{p_2 - p_1}{\gamma}$$
$$= \frac{(12,000 - 2000) \text{ lbf/ft}^2}{62.4 \text{ lbf/ft}^3}$$
$$= 160 \text{ ft}$$

The flow rate of water by weight is

$$\gamma Q = \gamma V A$$
$$= \left(62.4 \text{ lbf/ft}^2\right)(3 \text{ ft/s}) \left(\frac{\pi \times 0.3333^2}{4} \text{ ft}^2\right)$$
$$= 16.3 \text{ lbf/s}$$

Power is

$$P = h_p (\gamma Q)/\eta$$
$$= \frac{(160 \text{ ft}) \times (16.3 \text{ lbf/s})}{0.7}$$
$$= 3730 \text{ ft-lbf/s}$$

Converting units to horsepower

$$P = \frac{3730 \text{ ft-lbf/s}}{550 \text{ ft-lbf/(s-hp)}}$$
$$= \underline{\underline{6.78 \text{ hp}}}$$

## Problem 7.4

The following sketch shows a small, hand-held sprayer to be used by homeowners. Water flows through the 6-ft-long by 3/8-in.-diameter hose. The hose is terminated with a 1/16 in. diameter nozzle and the water exits the nozzle with a speed of 25 ft/s. Air in the tank is pressurized to produce the given flow. Determine the air pressure. Head loss in the system is given by $h_L = 5.0 \, (L/D) \, (V^2/2g)$, where $L = 6$ ft is the length of the hose, $D$ is the diameter of the hose, and $V$ is the average velocity in the hose. Assume all kinetic energy correction factors are unity.

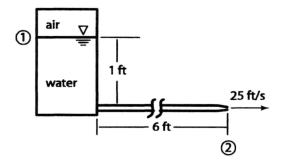

### Solution

From continuity, the water speed in the hose is

$$V_{\text{hose}} = V_{\text{nozzle}} \frac{(D_{\text{nozzle}})^2}{(D_{\text{hose}})^2}$$
$$= 25 \frac{\left(\frac{1}{16}\right)^2}{\left(\frac{3}{8}\right)^2}$$
$$= 0.694 \text{ ft/s}$$

The energy equation between section 1 and section 2 is

$$\frac{p_1}{\gamma} + \alpha_1 \frac{\overline{V}_1^2}{2g} + z_1 + h_p = \frac{p_2}{\gamma} + \alpha_2 \frac{\overline{V}_2^2}{2g} + z_2 + h_t + h_L$$

Now $\overline{V}_1 \approx 0$, $h_p = h_t = 0$, $\alpha_2 = 1$ and $p_2 = 0$. The energy equation simplifies to

$$\frac{p_{\text{air}}}{\gamma} + z_1 = \frac{V_{\text{nozzle}}^2}{2g} + z_2 + 5.0\,(L/D)\left(V_{\text{tube}}^2/2g\right)$$

Substituting values

$$\frac{p_{\text{air}}}{62.4} + 1 = \frac{25^2}{2 \times 32.2} + 0 + 5.0\,(6 \times 12/0.375)\left(\frac{0.694^2}{2 \times 32.2}\right)$$
$$= 16.88 \text{ ft}$$

So

$$p_{\text{air}} = 62.4\,(16.88 - 1)$$
$$= 991 \text{ lbf/ft}^2$$
$$= \underline{\underline{6.88 \text{ psi}}}$$

# Problem 7.5

Water from behind a dam flows through a turbine that is 85% efficient. The discharge is 12 m³/s, the head loss is 5 m, and all kinetic energy correction factors are unity. Determine the power output from the turbine.

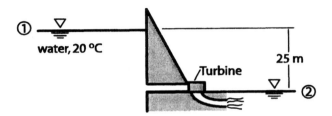

## Solution

The energy equation between sections 1 and 2 is

$$\frac{p_1}{\gamma} + \alpha_1 \frac{\overline{V}_1^2}{2g} + z_1 + h_p = \frac{p_2}{\gamma} + \alpha_2 \frac{\overline{V}_2^2}{2g} + z_2 + h_t + h_L$$

Now $p_1 = p_2 = 0, \overline{V}_1 \approx \overline{V}_2 \approx 0$, and $z_2 = 0$. The energy equation simplifies to

$$z_1 = h_t + h_L$$
$$25 = h_t + 5 \text{ m}$$

The power from the turbine is

$$P = \gamma Q h_t \eta$$
$$= 9810 \times 12 \times 20 \times 0.85$$
$$= 2000 \text{ kW}$$

## Problem 7.6

A centrifugal pump will be used to transport water at 50°F from a lake to a cabin. The discharge will be 10 gpm at an elevation of 60 ft above the lake surface. Discharge pressure is atmospheric. The suction pipe is 15 ft-long by 1-in.-diameter, and the discharge pipe is 200-ft-long by 1-in.-diameter. Head loss in each pipe is given by $h_L = 0.05\,(L/D)\,V^2/(2g)$, where $L$ and $D$ are pipe length and diameter, respectively. Assume the kinetic energy correction factor in each pipe is 1.0. Determine the head supplied by the pump, and sketch an energy and hydraulic grade line.

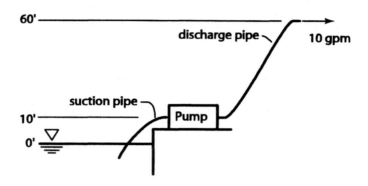

### Solution

The water speed in each pipe is

$$V = Q/A$$
$$= \frac{10 \times 0.002228}{\pi/4 \times (1/12)^2}$$
$$= 4.085 \text{ ft/s}$$

The energy equation between section 1 along the lake surface and section 2 at the discharge pipe exit is

$$\frac{p_1}{\gamma} + \alpha_1 \frac{\overline{V}_1^2}{2g} + z_1 + h_p = \frac{p_2}{\gamma} + \alpha_2 \frac{\overline{V}_2^2}{2g} + z_2 + h_t + h_L$$

Now $p_1 = p_2 = 0$, $\overline{V}_1 \approx 0$, $\alpha_2 = 1$, $z_2 = 60$, and $h_t = 0$. The energy equation simplifies to

$$h_p = \frac{\overline{V}_2^2}{2g} + z_2 + 0.05 \left(\frac{\overline{V}_2^2}{2g}\right)\left(\frac{L}{D}\right)_{\text{suction pipe}} + 0.05 \left(\frac{\overline{V}_2^2}{2g}\right)\left(\frac{L}{D}\right)_{\text{discharge pipe}}$$

Substituting values

$$h_p = \frac{4.075^2}{2 \times 32.2} + 60 + 0.05 \left(\frac{4.075^2}{2 \times 32.2}\right) \frac{15}{1/12} + 0.05 \left(\frac{4.075^2}{2 \times 32.2}\right) \frac{200}{1/12}$$
$$= (0.257 + 60 + 2.32 + 30.94) \text{ ft} \tag{1}$$
$$= \underline{93.5 \text{ ft}}$$

Prior to sketching the hydraulic grade line (HGL) and the energy grade line (EGL), notice that Eq. (1) provides the following values:

$$\text{velocity head in each pipe} = 0.257 \text{ ft}$$
$$h_L \text{ in suction pipe} = 2.32 \text{ ft}$$
$$h_L \text{ in discharge pipe} = 30.94 \text{ ft}$$

To develop the HGL and EGL plots, begin with the energy equation.

$$\frac{p_1}{\gamma} + \alpha_1 \frac{\overline{V}_1^2}{2g} + z_1 + h_p = \frac{p_2}{\gamma} + \alpha_2 \frac{\overline{V}_2^2}{2g} + z_2 + h_t + h_L$$

When $h_p = h_t = 0$, this can be written as

$$EGL(1) = EGL(2) + h_L(2) \tag{2}$$

where $EGL(1) = \frac{p_1}{\gamma} + \alpha_1 \frac{\overline{V}_1^2}{2g} + z_1$ and $EGL(2)$ has a similar definition. Define locations $a$ through $e$ as shown in Fig. 1.

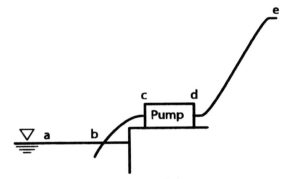

Figure 1 Sketch of the system

Between points $a$ and $b$, there is no head loss, and Eq. (2) simplifies to

$$EGL(a) = EGL(b)$$
$$= 0 \text{ ft}$$

where the value of 0 ft arises because the lake surface has a pressure of zero and an elevation of zero.

Let $x$ be an arbitrary location between $b$ and $c$. Eq. (2) becomes

$$EGL(b) = EGL(x) + h_L(x)$$

where $EGL(b) = 0$. Since $h_L$ is linear with $x$, it can be written as $h_L(x) = 2.32(x/15)$. Thus

$$EGL(x) = -2.32(x/15)$$
$$\text{for } b \leq x \leq c$$

Between $c$ and $d$, the pump adds 93.5 ft of head. Thus

$$EGL(d) = EGL(c) + 93.5$$
$$= -2.32 + 93.5$$
$$= 91.2 \text{ ft}$$

Between $d$ and $e$, the $EGL$ is given by

$$EGL(x) = 91.2 - 30.9(x/200)$$
$$\text{for } d \leq x \leq e$$

where 30.9 ft is the head loss in the discharge pipe and 200 ft is the length of the discharge pipe. From this equation, $EGL(e) = 60.3$ ft.

A plot of the EGL is shown in Fig. 2.

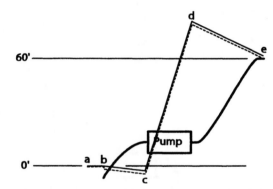

Figure 2 Energy grade line (EGL) and hydraulic grade line (HGL)

Fig. 2 also shows the HGL. The HGL was found by subtracting the velocity head (0.26 ft) from the EGL.

# Chapter 8

# Dimensional Analysis and Similitude

## Problem 8.1

The discharge, $Q$, of an ideal fluid (no viscous effects) through an orifice depends on the orifice diameter, $d$, the pressure drop across the orifice, $\Delta p$, and the fluid density. Find a nondimensional relationship for the discharge.

### Solution

The functional relationship is

$$Q = f(d, \Delta p, \rho)$$

Based on the Buckingham $\Pi$ theorem, there should be $4 - 3 = 1$ $\pi$–groups. Use the step-by-step method, as shown in the following table.

| Variable | [ ] | Variable | [ ] | Variable | [ ] | Variable | [ ] |
|---|---|---|---|---|---|---|---|
| $Q$ | $\frac{L^3}{T}$ | $\frac{Q}{d^3}$ | $\frac{1}{T}$ | $\frac{Q}{d^3}$ | $\frac{1}{T}$ | $\frac{Q}{d^2}\sqrt{\frac{\rho}{\Delta p}}$ | 0 |
| $d$ | $L$ | | | | | | |
| $\rho$ | $\frac{M}{L^3}$ | $\rho d^3$ | $M$ | | | | |
| $\Delta p$ | $\frac{M}{LT^2}$ | $\Delta p d$ | $\frac{M}{T^2}$ | $\frac{\Delta p}{\rho d^2}$ | $\frac{1}{T^2}$ | | |

As shown in the table, the length dimension is first eliminated with $d$, then the mass dimension is eliminated with $\rho d^3$, and finally the time dimension is eliminated with $\Delta p/\rho d^2$. Thus

$$\frac{Q}{d^2}\sqrt{\frac{\rho}{\Delta p}} = K$$

where $K$ is a constant. This may be expressed as

$$Q = Kd^2\sqrt{\frac{\Delta p}{\rho}}$$

# Problem 8.2

The terminal velocity of a sphere (maximum drop velocity) depends on the sphere diameter, sphere density, fluid density, fluid viscosity, and acceleration due to gravity.

$$V_t = f(D, \rho_s, \rho_f, \mu, g)$$

Find a nondimensional form for the terminal velocity.

## Solution

Based on the Buckingham Π theorem, there should be $6 - 3 = 3$ $\pi$-groups. Use the step-by-step method as shown in the following table.

| Variable | [ ] | Variable | [ ] | Variable | [ ] | Variable | [ ] |
|---|---|---|---|---|---|---|---|
| $V_t$ | $\frac{L}{T}$ | $\frac{V_t}{D}$ | $\frac{1}{T}$ | $\frac{V_t}{D}$ | $\frac{1}{T}$ | $\frac{V_t \rho_f D}{\mu}$ | 0 |
| $\rho_s$ | $\frac{M}{L^3}$ | $\rho_s D^3$ | $M$ | $\frac{\rho_s}{\rho_f}$ | 0 | | |
| $\rho_f$ | $\frac{M}{L^3}$ | $\rho_f D^3$ | $M$ | | | | |
| $\mu$ | $\frac{M}{LT}$ | $\mu D$ | $\frac{M}{T}$ | $\frac{\mu}{\rho_f D^2}$ | $\frac{1}{T}$ | | |
| $D$ | $L$ | | | | | | |
| $g$ | $\frac{L}{T^2}$ | $\frac{g}{D}$ | $\frac{1}{T^2}$ | $\frac{g}{D}$ | $\frac{1}{T^2}$ | $\frac{g\rho_f^2 D^3}{\mu^2}$ | 0 |

First, length is eliminated with $D$, then mass is eliminated with $\rho_f D^3$, and finally time is eliminated with $\mu/\rho_f D^2$.

The nondimensional grouping becomes

$$\frac{V_t \rho_f D}{\mu} = f\left(\frac{\rho_s}{\rho_f}, \frac{g\rho_f^2 D^3}{\mu^2}\right)$$

which can also be written as

$$V_t = \frac{\mu}{\rho_f D} f\left(\frac{\rho_s}{\rho_f}, \frac{\sqrt{gD}\rho_f D}{\mu}\right)$$

or

$$V_t = \sqrt{gD} f\left(\frac{\rho_s}{\rho_f}, \frac{\sqrt{gD}\rho_f D}{\mu}\right)$$

## Problem 8.3

The pressure drop in a smooth horizontal pipe in a turbulent, incompressible flow depends on the pipe diameter, pipe length, fluid velocity, fluid density, and viscosity.

$$\Delta p = f(D, S, V, \rho, \mu)$$

Find a nondimensional relationship for the pressure drop.

## Solution

By the Buckingham $\Pi$ theorem, the number of dimensionless $\pi$-groups is $6 - 3 = 3$. The exponent method will be used. First, express the equation as

$$\Delta p = D^\alpha S^\beta V^\gamma \rho^\delta \mu^\epsilon$$

Substitute the dimensions of each variable

$$\frac{M}{LT^2} = L^\alpha L^\beta \left(\frac{L}{T}\right)^\gamma \left(\frac{M}{L^3}\right)^\delta \left(\frac{M}{LT}\right)^\epsilon$$

Equate the powers of each dimension

$$\begin{aligned} M: & \quad 1 = \delta + \epsilon \\ L: & \quad -1 = \alpha + \beta + \gamma - 3\delta - \epsilon \\ T: & \quad -2 = -\gamma - \epsilon \end{aligned}$$

Solving for $\alpha$, $\gamma$ and $\delta$ in terms of $\beta$ and $\epsilon$

$$\begin{aligned} \delta &= 1 - \epsilon \\ \gamma &= 2 - \epsilon \\ \alpha &= -\beta - \epsilon \end{aligned}$$

Substituting back into the equation for pressure

$$\Delta p = D^{-\beta-\epsilon} S^\beta V^{2-\epsilon} \rho^{1-\epsilon} \mu^\epsilon$$

$$= \rho V^2 \left(\frac{S}{D}\right)^\beta \left(\frac{\mu}{VD\rho}\right)^\epsilon$$

This relation can be expressed as

$$\underline{\underline{\frac{\Delta p}{\rho V^2} = f\left(\frac{S}{D}, \frac{\rho V D}{\mu}\right)}}$$

## Problem 8.4

A $\frac{1}{25}$ scale model of an airship is tested in water at 20°C. If the airship travels 5 m/s in air at atmospheric pressure and 20°C, find the velocity for the model to achieve similitude. Also, find the ratio of the drag force on the prototype to that on the model. The densities of water and air at these conditions are 1000 kg/m³ and 1.2 kg/m³. The corresponding dynamic viscosities of water and air are $10^{-3}$ N·s/m² and $1.81 \times 10^{-5}$ N·s/m².

### Solution

The significant nondimensional number for this problem is the Reynolds number. Thus, for similitude

$$\text{Re}_{\text{model}} = \text{Re}_{\text{prototype}}$$

$$\frac{V_m L_m \rho_m}{\mu_m} = \frac{V_p L_p \rho_p}{\mu_p}$$

or

$$V_m = V_p \frac{L_p}{L_m} \frac{\rho_p}{\rho_m} \frac{\mu_m}{\mu_p}$$

$$= 5 \times 25 \times \frac{1.2}{1000} \frac{10^{-3}}{1.81 \times 10^{-5}}$$

$$= \underline{\underline{8.29 \text{ m/s}}}$$

Dimensional analysis for the force yields

$$F = \rho V^2 L^2 f(\text{Re})$$

Thus, for the ratio of forces

$$\frac{F_p}{F_m} = \frac{\rho_p}{\rho_m} \frac{V_p^2}{V_m^2} \frac{L_p^2}{L_m^2} \frac{f(\text{Re}_p)}{f(\text{Re}_m)}$$

Since the Reynolds numbers are the same

$$\frac{F_p}{F_m} = \frac{\rho_p}{\rho_m} \frac{V_p^2}{V_m^2} \frac{L_p^2}{L_m^2}$$

The force ratio is

$$\frac{F_p}{F_m} = \frac{1.2}{1000} \frac{5^2}{8.29^2} 25^2$$

$$= \underline{\underline{0.273}}$$

## Problem 8.5

A scale model of a pumping system is to be tested to determine the head losses in the actual system. Air with a specific weight of 0.085 lbf/ft$^3$ and a viscosity of $3.74 \times 10^{-7}$ lbf·s/ft$^2$ is to be used in the model. Water with a specific weight of 62.4 lbf/ft$^3$ and a viscosity of $2.36 \times 10^{-5}$ lbf-s/ft$^2$ is used in the prototype. The velocity in the prototype is 2 ft/s. A practical upper limit for the air velocity in the model to avoid compressibility effects is 100 ft/s. Find the scale ratio for the model and the ratio of the pressure losses in the prototype to those in the model.

### Solution

In this problem, the Reynolds number is the important scaling parameter so

$$\text{Re}_{\text{model}} = \text{Re}_{\text{prototype}}$$
$$\frac{V_m L_m \rho_m}{\mu_m} = \frac{V_p L_p \rho_p}{\mu_p}$$

Therefore

$$\frac{L_m}{L_p} = \frac{V_p \rho_p \mu_m}{V_m \rho_m \mu_p}$$
$$= \frac{2}{100} \frac{(62.4/32.2)}{(0.085/32.2)} \frac{3.74 \times 10^{-7}}{2.36 \times 10^{-5}}$$
$$= \underline{0.233}$$

or about a $\frac{1}{4}$ scale model. Note that the specific weight is changed to mass density by dividing by 32.2.

Since the Reynolds numbers are the same, the pressure coefficients are also the same.

$$C_{p,m} = C_{p,p}$$
$$\left(\frac{\Delta p}{\rho V^2}\right)_m = \left(\frac{\Delta p}{\rho V^2}\right)_p$$

or

$$\frac{\Delta p_p}{\Delta p_m} = \frac{\rho_p}{\rho_m} \frac{V_p^2}{V_m^2}$$

which gives

$$\frac{\Delta p_p}{\Delta p_m} = \frac{(62.4/32.2)}{(0.085/32.2)} \frac{2^2}{100^2}$$
$$= \underline{0.294}$$

## Problem 8.6

The sloshing of oil in a tank is affected by both viscous and gravitational effects. A 1:4 scale model of oil with a kinematic viscosity of $1.1 \times 10^{-4}$ m$^2$/s is to be used to study the sloshing. Find the kinematic viscosity of the liquid to be used in the model.

## Solution

In this problem, both the Reynolds number and Froude number need to be the same. For Froude number scaling

$$Fr_m = Fr_p$$
$$\frac{gL_m}{V_m^2} = \frac{gL_p}{V_p^2}$$

so

$$\frac{V_m}{V_p} = \sqrt{\frac{L_m}{L_p}}$$
$$= \sqrt{\frac{1}{4}} = \frac{1}{2}$$

For Reynolds number scaling

$$\text{Re}_{\text{model}} = \text{Re}_{\text{prototype}}$$
$$\frac{V_m L_m}{\nu_m} = \frac{V_p L_p}{\nu_p}$$

so

$$\nu_m = \nu_p \frac{V_m}{V_p} \frac{L_m}{L_p}$$
$$= 1.1 \times 10^{-4} \frac{1}{2} \frac{1}{4}$$
$$= \underline{\underline{1.37 \times 10^{-5} \text{ m}^2/\text{s}}}$$

# Problem 8.7

A wind-tunnel test is performed on a $\frac{1}{20}$ scale model of a supersonic aircraft. The prototype aircraft flies at 480 m/s in conditions where the speed of sound is 300 m/s and the air density is 1.0 kg/m$^3$. The model aircraft is tested in a wind tunnel in which the speed of sound is 279 m/s and the air density is 0.43 kg/m$^3$. The drag force on the model is 100 N. What speed must the flow in the wind tunnel be for dynamic similitude, and what is the drag force on the prototype?

## Solution

The primary dimensionless number is the Mach number.

$$M_m = M_p$$
$$\frac{V_m}{c_m} = \frac{V_p}{c_p}$$

Thus

$$\begin{aligned} V_m &= V_p \frac{c_m}{c_p} \\ &= 480 \frac{279}{300} \\ &= 446 \text{ m/s} \end{aligned}$$

The nondimensional form for the drag force is

$$\frac{D}{\rho V^2 L^2} = f(M)$$

so

$$\begin{aligned} \frac{D_p}{D_m} &= \frac{\rho_p V_p^2}{\rho_m V_m^2} \frac{L_p^2}{L_m^2} \\ &= \frac{1 \times 480^2}{0.43 \times 446^2} \times 20^2 \\ &= 1077 \end{aligned}$$

The drag on the prototype is

$$D_p = 1077 \times 100 = \underline{108 \text{ kN}}$$

## Problem 8.8

The surface tension of pure water is 0.073 N/m, and the surface tension of soapy water is 0.025 N/m. If a pure water droplet breaks up in an airstream that is moving at 10 m/s, at what speed would the same size soapy-water droplet break up?

### Solution

The significant dimensionless parameter for droplet breakup is the Weber number. It is assumed that breakup will occur at the same Weber numbers. The Weber number is

$$W = \frac{\rho V^2 L}{\sigma}$$

In this case, the dimension $L$ is the droplet diameter.

$$\left(\frac{\rho V^2 D}{\sigma}\right)_{\text{soapy}} = \left(\frac{\rho V^2 D}{\sigma}\right)_{\text{pure}}$$

Since the density and diameter are the same on both sides of the equation

$$V_{\text{soapy}} = V_{\text{pure}} \sqrt{\frac{\sigma_{\text{soapy}}}{\sigma_{\text{pure}}}}$$
$$= 10 \times \sqrt{\frac{0.025}{0.073}}$$
$$= \underline{\underline{5.85 \text{ m/s}}}$$

## Problem 8.9

A 1:49 scale model of a ship is tested in a water tank. The speed of the prototype is 10 m/s. The purpose of the tests is to measure the wave drag on the ship. Find the velocity of the model and the ratio of the wave drag on the prototype to that on the model.

### Solution

The wave drag is due to gravitational effects, so Froude number scaling is used.

$$Fr_m = Fr_p$$
$$\left(\frac{V^2}{gL}\right)_m = \left(\frac{V^2}{gL}\right)_p$$

Thus
$$V_m = V_p\sqrt{\frac{L_m}{L_p}}$$
$$V_m = 10 \times \sqrt{\frac{1}{49}}$$
$$= \underline{\underline{1.43 \text{ m/s}}}$$

The nondimensional form for the wave drag is
$$\frac{D}{\rho V^2 L^2} = f(Fr)$$

Because the Froude numbers are the same
$$\frac{D_p}{D_m} = \frac{\rho_p V_p^2}{\rho_m V_m^2}\frac{L_p^2}{L_m^2}$$

The density for the prototype and model are the same, so
$$\frac{D_p}{D_m} = \frac{49}{1}\frac{49^2}{1}$$
$$= \underline{\underline{1.18 \times 10^5}}$$

# Chapter 9

# Surface Resistance

## Problem 9.1

An aluminum cube of density 2700 kg/m³ slides with a constant speed of 20 cm/s down a plate that is at an angle of 30° with respect to the horizontal. The plate is covered with a stationary layer of 0.1-mm-thick oil of viscosity $\mu = 0.008$ N·s/m². The cube has dimensions of $L \times L \times L$. Find $L$.

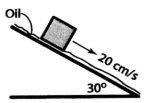

### Solution

The weight of the cube is
$$W = L^3 \gamma_{Al}$$

A free-body diagram is

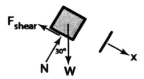

Balancing forces in the $x$-direction gives
$$F_{\text{shear}} = W \sin(30°)$$
$$= (L^3 \gamma_{Al}) \sin(30°) \qquad (1)$$

The shear force is

$$F_{\text{shear}} = \tau L^2 \qquad (2)$$

Assuming Couette flow, the shear stress is

$$\tau = \mu \frac{V}{h} \qquad (3)$$

where $V$ is the speed of the block and $h$ is the thickness of the oil layer. Combining Eqs. (2) and (3) gives

$$F_{\text{shear}} = \left(\mu \frac{V}{h}\right) L^2 \qquad (4)$$

Combining Eqs. (1) and (4) gives

$$\left(\mu \frac{V}{h}\right) L^2 = (L^3 \gamma_{Al}) \sin(30^\circ)$$

or

$$L = \mu \frac{V}{h(\gamma_{Al} \sin 30^\circ)}$$
$$= 0.008 \frac{0.2}{0.0001\,(2700 \times \sin 30)}$$
$$= \underline{11.9 \text{ mm}}$$

# Problem 9.2

A 1.000-in.-diameter shaft of length 2 inches rotates at an angular speed of $\omega = 800$ rpm within a stationary cylindrical housing. The gap between the stationary housing and the shaft has a dimension of 0.001 in. The gap is filled with oil of viscosity 0.0003 lbf·s/ft$^2$. Find the torque and power required to rotate the shaft. Assume the oil motion in the gap can be described by planar Couette flow.

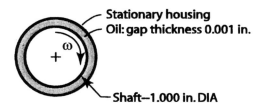

## Solution

As the shaft rotates, a clockwise applied torque is required to balance the moment caused by forces associated with fluid friction (viscosity).

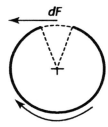

Figure 1  Sketch of the shaft

The force $dF$ in Fig. 1 is the viscous force on the top surface of the pie-shaped region. Since shear stress is force per area

$$dF = \tau dA$$
$$= \tau (rd\theta) L \qquad (1)$$

The shear stress is

$$\tau = \mu \frac{dV}{dy}$$

which simplifies, for Couette flow, to

$$\tau = \mu \frac{V}{h}$$

where $V = r\omega$ is the speed at the outer surface of the rotating shaft. Thus

$$\tau = \mu \frac{r\omega}{h} \qquad (2)$$

Combining Eqs. (1) and (2) gives

$$dF = \mu \frac{r\omega}{h} (rd\theta) L$$

The force $dF$ will cause a frictional torque of $dT$.

$$dT = rdF$$
$$= \mu \frac{r^3 L\omega}{h} d\theta$$

The net frictional torque balances the applied torque.

Applied torque = frictional torque
$$= \int_0^{2\pi} \mu \frac{r^3 L\omega}{h} d\theta$$
$$= 2\pi\mu \frac{r^3 L\omega}{h}$$
$$= 2\pi \left(0.0003 \frac{\text{lbf} \cdot \text{s}}{\text{ft}^2}\right) \left(\frac{0.5^3 \times 2}{0.001} \text{in.}^3\right) \left(\frac{\text{ft}^3}{12^3 \text{ in.}^3}\right) \left(\frac{2\pi \times 800}{60} \frac{1}{\text{s}}\right)$$
$$= \underline{0.0228 \text{ ft-lbf}}$$

Power $P$ is

$$P = (\text{applied torque})(\text{angular speed})$$
$$= (0.0228 \text{ ft-lbf}) \left(\frac{2\pi \times 800}{60} \frac{1}{\text{s}}\right) \left(\frac{\text{hp-s}}{550 \text{ ft-lbf}}\right)$$
$$= 0.00347 \text{ hp}$$

# Problem 9.3

Oil with viscosity 0.0014 lbf-s/ft² and density 1.71 slug/ft³ flows between two parallel plates that are spaced 0.125 in. apart and inclined at a 45° angle. Pressure gages at locations $A$ and $B$ indicate that $p_A - p_B = 5$ psi. The distance between the pressure gages is 2 ft. Each plate has a dimension (i.e., depth into the paper) of 1.5 ft. Determine the rate of volume flow of oil.

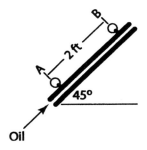

## Solution

Laminar flow between parallel plates (planar Poiseuille flow) is described by:.

$$-\left(\frac{12\mu}{B^3}\right) q = \frac{dp}{ds} + \gamma \frac{dz}{ds} \qquad (1)$$

The left side of Eq. (1) is

$$-\left(\frac{12\mu}{B^3}\right) q = -\left(\frac{12 \times 0.0014 \text{ lbf-s/ft}^2}{(0.125/12)^3 \text{ ft}^3}\right) q$$

$$= \left(-14,864 \text{ lbf-s/ft}^5\right) q$$

The pressure gradient is

$$\frac{dp}{ds} = \frac{\Delta p}{\Delta s}$$

$$= \left(\frac{-5 \text{ lbf/in.}^2}{2 \text{ ft}}\right) \left(\frac{144 \text{ in.}^2}{\text{ft}^2}\right)$$

$$= -360 \text{ lbf/ft}^3$$

The slope term is

$$\gamma \frac{dz}{ds} = \gamma \sin(45°)$$

$$= 1.71 \times 32.2 \times \sin(45°)$$

$$= 38.93 \text{ lbf/ft}^3$$

Substituting numerical values into Eq. (1) gives

$$-\left(\frac{12\mu}{B^3}\right)q = \frac{dp}{ds} + \gamma\frac{dz}{ds}$$

$$\left(-14,864 \text{ lbf-s/ft}^5\right)q = \left(-360 \text{ lbf/ft}^3\right) + \left(38.93 \text{ lbf/ft}^3\right)$$

So

$$q = \frac{-360 + 38.93}{-14,864}$$
$$= 0.0216 \text{ ft}^2/\text{s}$$

The volume rate of flow is

$$Q = qw$$
$$= (0.0216 \text{ ft}^2/\text{s})(1.5 \text{ ft})$$
$$= \underline{\underline{0.0324 \text{ ft}^3/\text{s}}}$$

Checking the Reynolds number

$$\text{Re} = \frac{q\rho}{\mu}$$
$$= \frac{(0.0216 \text{ ft}^2/\text{s})(1.71 \text{ slug/ft}^3)}{0.0014 \text{ lbf-s/ft}^2}$$
$$= 26.4$$

Since this is far less than the critical value of the Reynolds number for turbulent flow (1000), the flow is laminar and Eq. (9.1) is valid.

## Problem 9.4

A thin plate that is 75 cm long and 30 cm wide is submerged and held stationary in a stream of water ($T = 10\,°C$) that has a speed of 2 m/s. What is the thickness of the boundary layer on the plate at the location where $\text{Re}_x = 500,000$ and at what distance $x$ does this Reynolds number occur? What is the shear stress on the plate at this point?

### Solution

The Reynolds number is

$$500,000 = \frac{\rho V x}{\mu}$$

$$500,000 = \frac{1000 \times 2 \times x}{1.31 \times 10^{-3}}$$

So, the distance $x$ is

$$x = \underline{32.8 \text{ cm}}$$

The boundary layer thickness is

$$\delta = \frac{5x}{\sqrt{\text{Re}_x}}$$
$$= \frac{5 \times 32.8 \text{ cm}}{\sqrt{500,000}}$$
$$= \underline{2.32 \text{ mm}}$$

The local shear stress coefficient is

$$c_f = \frac{0.664}{\sqrt{\text{Re}_x}}$$
$$= \frac{0.664}{\sqrt{500,000}}$$
$$= 9.39 \times 10^{-4}$$

The local wall shear stress is

$$\tau_o = c_f \left(\frac{\rho U_o^2}{2}\right)$$
$$= 9.39 \times 10^{-4} \left(\frac{1000 \times 2^2}{2}\right)$$
$$= \underline{1.88 \text{ Pa}}$$

# Problem 9.5

Air with a kinematic viscosity of $15.1 \times 10^{-6}$ m$^2$/s, a density of 1.2 kg/m$^3$ and a free stream velocity of 30 m/s flows over a 0.8-m-long by 0.2-m wide flat plate. Find the wall shear stress at a horizontal distance of 0.5 m. Also, find the shear force on the top side of the plate.

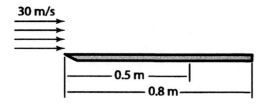

## Solution

To find the wall shear stress, begin by finding the local Reynolds number at $x = 0.5$ m.

$$\text{Re}_x = \frac{Vx}{\nu}$$
$$= \frac{30 \times 0.5}{15.1 \times 10^{-6}}$$
$$= 993{,}400$$

Since $\text{Re}_x > 500{,}000$, the boundary layer at this location is turbulent. The local shear stress coefficient is

$$c_f = \frac{0.455}{\ln^2(0.06\,\text{Re}_x)}$$
$$= \frac{0.455}{\ln^2(0.06 \times 993{,}400)}$$
$$= 0.003763$$

The wall shear stress is

$$\tau_o = c_f \left(\frac{\rho U_o^2}{2}\right)$$
$$= 0.003763 \left(\frac{1.2 \times 30^2}{2}\right)$$
$$= 2.03 \text{ Pa}$$

To find the shear force on the plate, begin by finding the Reynolds number based on plate length.

$$\text{Re}_L = \frac{VL}{\nu}$$
$$= \frac{30 \times 0.8}{15.1 \times 10^{-6}}$$
$$= 1,589,000$$

The average shear stress coefficient is

$$C_f = \frac{0.523}{\ln^2(0.06\,\text{Re}_L)} - \frac{1520}{\text{Re}_L}$$
$$= \frac{0.523}{\ln^2(0.06 \times 1,589,000)} - \frac{1520}{1,589,000}$$
$$= 0.003022$$

The shear force is

$$F_s = C_f \left(\frac{\rho U_o^2}{2}\right) Lw$$
$$= 0.003022 \left(\frac{1.2 \times 30^2}{2}\right)(0.8 \times 0.2)$$
$$= \underline{0.261\ \text{N}}$$

## Problem 9.6

Assuming that drag is entirely due to *skin-friction drag*, find the drag force and power for a person swimming. Assume that the human body can be represented as a submerged, thin, flat plate of dimension (30 cm) × (180 cm) with drag occurring on both sides of the plate. Use a swimming speed of 1.5 m/s, a water density of 1000 kg/m³ and a dynamic viscosity of 0.00131 N-s/m².

### Solution

Drag force $F_D$ is

$$F_D = C_f \left(\frac{\rho U_o^2}{2}\right) 2Lw \tag{1}$$

To find the average skin friction coefficient, the value of $\text{Re}_L$ is needed.

$$\text{Re}_L = \frac{VL}{\nu}$$
$$= \frac{1.5 \times 1.8}{1.31 \times 10^{-6}}$$
$$= 2,061,000$$

Since this Reynolds number is above 500,000, the boundary layer is mixed (laminar followed by turbulent).

The average skin friction coefficient is

$$C_f = \frac{0.523}{\ln^2(0.06\,\text{Re}_L)} - \frac{1520}{\text{Re}_L}$$
$$= \frac{0.523}{\ln^2(0.06 \times 2{,}061{,}000)} - \frac{1520}{2{,}061{,}000}$$
$$= 0.00307$$

Substituting into Eq. (1) gives

$$F_D = C_f \left(\frac{\rho U_o^2}{2}\right)(2Lw)$$
$$= 0.00307 \left(\frac{1000 \times 1.5^2}{2}\right)(2 \times 1.8 \times 0.3)$$
$$= \underline{3.73\ \text{N}}$$

Power to overcome surface drag is the product of force and speed.

$$P = F_D U_o$$
$$= (3.73\ \text{N})(1.5\ \text{m/s})$$
$$= \underline{5.60\ \text{W}}$$

# Problem 9.7

The small toy airplane that is shown in the following photo is flying with a speed of 2.5 m/s. Represent the wing as a thin, flat plate of dimension 5-cm × 30-cm, and find the drag force on the wing. Also, determine the power that must be supplied by the propeller for constant-speed, level flight. For this calculation, assume that the shear force on the wing is 50% of the total drag force. For air, use a kinematic viscosity of $15.1 \times 10^{-6}$ m$^2$/s and a density of 1.2 kg/m$^3$.

## Solution

To find the shear force on the wing, begin by finding the Reynolds number based on cord.

$$\begin{aligned}
\mathrm{Re}_L &= \frac{VL}{\nu} \\
&= \frac{2.5 \times 0.05}{15.1 \times 10^{-6}} \\
&= 8,278
\end{aligned}$$

Thus the boundary layer is laminar, and the average shear stress coefficient is

$$\begin{aligned}
C_f &= \frac{1.33}{\mathrm{Re}_L^{1/2}} \\
&= \frac{1.33}{8,278^{1/2}} \\
&= 0.01462
\end{aligned}$$

The shear force is

$$F_s = C_f \left(\frac{\rho U_o^2}{2}\right) 2Lw$$
$$= 0.01462 \left(\frac{1.2 \times 2.5^2}{2}\right) 2(0.3 \times 0.05)$$
$$= \underline{\underline{1.64 \times 10^{-3} \text{ N}}}$$

To find the power output of the propeller, begin with a free-body diagram.

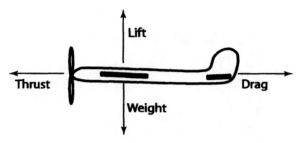

In the horizontal direction, the thrust of the propeller exactly balances drag. The power ($P$) produced by the propeller is given by the product of speed ($V$) and thrust force ($F_T$).

$$P = F_T V$$
$$= F_D V$$
$$= \left(\frac{1.64 \times 10^{-3} \text{ N}}{0.5}\right)\left(\frac{2.5 \text{ m}}{\text{s}}\right)$$
$$= \underline{\underline{8.2 \text{ mW}}}$$

# Chapter 10

# Flow in Conduits

## Problem 10.1

Water at 20°C ($\mu = 10^{-3}$ N·s/m$^2$, $\rho = 1000$ kg/m$^3$) flows through a 0.5-mm tube connected to the bottom of a reservoir. The length of the tube is 1.0 m, and the depth of water in the reservoir is 20 cm. Find the flow rate in the tube. Neglect the entrance loss at the junction of the tube and reservoir.

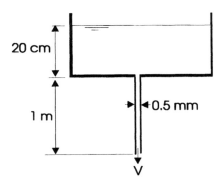

## Solution

Applying the energy equation between the top of the water in the reservoir (1) and the end of the tube (2) gives

$$\frac{p_1}{\gamma} + \alpha_1 \frac{V_1^2}{2g} + z_1 + h_p = \frac{p_2}{\gamma} + \alpha_2 \frac{V_2^2}{2g} + z_2 + h_t + h_L$$

The pressure at points 1 and 2 is the same (atmospheric), the velocity in the reservoir is zero, and there is no pump or turbine in the system. Also, the only losses are friction losses in the tube. The energy equation simplifies to

$$z_1 = \alpha_2 \frac{V_2^2}{2g} + z_2 + h_f$$

We will assume the flow is laminar, so $\alpha_2 = 2$. The head loss due to friction in a laminar flow is

$$h_f = 32 \frac{\mu L V}{\gamma D^2}$$

Substituting into the energy equation gives

$$z_1 = \alpha_2 \frac{V_2^2}{2g} + z_2 + 32 \frac{\mu L V_2}{\gamma D^2}$$

and replacing the variables with values

$$1.2 = \frac{V_2^2}{9.81} + 32 \frac{10^{-3} \times 1 \times V_2}{9810 \times 0.0005^2}$$

$$0.102 V_2^2 + 13.05 V_2 - 1.2 = 0$$

Solving

$$V_2 = 0.092 \text{ m/s}$$

The volume flow rate is

$$Q = AV = \frac{\pi}{4} \times 0.0005^2 \times 0.092 = 1.8 \times 10^{-8} \text{ m}^3/\text{s}$$
$$= \underline{\underline{1.8 \times 10^{-2} \text{ ml/s}}}$$

To determine whether the flow is laminar, calculate the Reynolds number.

$$\text{Re} = \frac{\rho V D}{\mu}$$
$$= \frac{10^3 \times 0.092 \times 0.0005}{10^{-3}}$$
$$= 46$$

Since the Reynolds number is less than 2000, the laminar flow assumption is justified.

# Problem 10.2

An oil supply line for a bearing is being designed. The supply line is a tube with an internal diameter of $\frac{1}{16}$ in. and 10 feet long. It is to transport SAE 30W oil ($\mu = 2 \times 10^{-3}$ lbf·s/ft², $\rho = 1.71$ slugs/ft³) at the rate of 0.01 gpm. Find the pressure drop across the line.

## Solution

First determine the Reynolds number to establish if the flow is laminar or turbulent. The velocity in the line is

$$V = \frac{Q}{A}$$
$$= \frac{0.01 \text{ gpm} \times 0.00223 \text{ cfs/gpm}}{\frac{\pi}{4} \times \left(\frac{1}{16} \times \frac{1}{12}\right)^2 \text{ ft}^2}$$
$$= 1.05 \text{ ft/s}$$

The Reynolds number is

$$\text{Re} = \frac{\rho V D}{\mu}$$
$$= \frac{1.71 \times 1.05 \times \frac{1}{16} \times \frac{1}{12}}{2 \times 10^{-3}}$$
$$= 4.68$$

The flow is laminar, so the head loss is

$$h_f = 32 \frac{\mu L V}{\gamma D^2}$$
$$= 32 \frac{2 \times 10^{-3} \times 10 \times 1.05}{1.71 \times 32.2 \times \left(\frac{1}{16} \times \frac{1}{12}\right)^2} = 450 \text{ ft}$$

The pressure drop is

$$\Delta p = \gamma h_f$$
$$= 1.71 \times 32.2 \times 450$$
$$= 24.8 \times 10^3 \text{ psf} = \underline{\underline{172 \text{ psi}}}$$

# Problem 10.3

Kerosene ($\mu = 1.9 \times 10^{-3}$ N·s/m$^2$, $S=0.81$) flows in a 2-cm diameter commercial steel pipe ($k_s = 0.046$ mm). A mercury manometer ($S = 13.6$) is connected between a 2-m section of pipe as shown, and there is a 5-cm deflection in the manometer. The elevation difference between the two taps is 0.5 mm. Find the direction and velocity of the flow in the pipe.

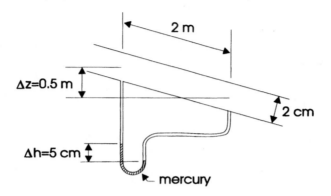

## Solution

First find the difference in piezometric pressure between the two pressure taps in the pipe. Flow is always in the direction of decreasing piezometric head. Take station 1 on the left and station 2 on the right. Define the distance $s$ as the distance from the center of the pipe at station 2 and the top of the mercury in the manometer. Using the manometer equation from 1 to 2 gives

$$p_2 = p_1 + \gamma_k(z_1 - z_2) + \gamma_k s + \gamma_{Hg}\Delta h - \gamma_k \Delta h - \gamma_k s$$

Thus we can write

$$p_2 + \gamma_k z_2 - (p_1 + \gamma_k z_1) = (\gamma_{Hg} - \gamma_k)\Delta h$$
$$p_{z,2} - p_{z,1} = (\gamma_{Hg} - \gamma_k)\Delta h$$

Since $\gamma_{Hg} > \gamma_k$, $p_{z,2} > p_{z,1}$, the flow must be from right to left (uphill).

The energy equation from 2 to 1 is

$$p_{z,2} + \alpha_2 \rho_k \frac{V_2^2}{2} + \gamma_k h_p = p_{z,1} + \alpha_1 \rho_k \frac{V_1^2}{2} + \gamma_k h_t + \gamma_k h_L$$

Since the pipe has a constant area, $V_1 = V_2$, and there are no turbines or pumps in the system, the equation reduces to

$$p_{z,2} - p_{z,1} = \gamma_k h_L = (\gamma_{Hg} - \gamma_k)\Delta h$$

The head loss can be expressed using the Darcy-Weisbach relation

$$h_L = f \frac{L}{D} \frac{V^2}{2g}$$

so

$$f \frac{L}{D} \frac{V^2}{2g} = \left(\frac{\gamma_{Hg}}{\gamma_k} - 1\right) \Delta h$$

Substituting in the values

$$0.05 \times \left(\frac{13.6}{0.81} - 1\right) = fV^2 \frac{2}{0.02} \frac{1}{2 \times 9.81}$$
$$fV^2 = 0.155 \text{ m}^2/\text{s}^2$$

Since $f$ depends on the Reynolds number (and velocity), this equation has to be solved by iteration. The relative roughness for the pipe is

$$\frac{k_s}{D} = \frac{0.046}{20} = 0.0023$$

From the Moody diagram (Fig. 10.8), the friction factor for a fully rough pipe would be about 0.025. This would give a velocity of

$$V = \sqrt{\frac{0.155}{0.025}} = 2.49 \text{ m/s}$$

The corresponding Reynolds number is

$$\text{Re} = \frac{\rho V D}{\mu} = \frac{1000 \times 0.81 \times 2.49 \times 0.02}{1.9 \times 10^{-3}}$$
$$= 2.12 \times 10^4$$

From the Moody diagram, the friction factor for this Reynolds number is 0.0305. The velocity is corrected to

$$V = \sqrt{\frac{0.155}{0.0305}} = 2.25 \text{ m/s}$$

The new Reynolds number is $1.92 \times 10^4$. The friction factor is 0.030, giving a velocity of 2.27 m/s. Further iterations would not significantly change the value so

$$V = \underline{\underline{2.27 \text{ m/s}}}$$

and the flow is from right to left.

# Problem 10.4

A pump is to be used to transfer crude oil ($\mu = 2 \times 10^{-4}$ lbf-s/ft$^2$, $S = 0.86$) from the lower tank to the upper tank at a flow rate of 100 gpm. The loss coefficient for the check valve is 5.0. The loss coefficients for the elbow and the inlet are 0.9 and 0.5, respectively. The 2-in. pipe is made from commercial steel ($k_s = 0.002$ in.) and is 40 ft long. The elevation distance between the liquid surfaces in the tanks is 10 ft. The pump efficiency is 80%. Find the power required to operate the pump.

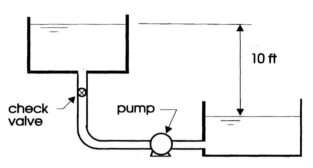

## Solution

To find the power required, we need to calculate the head provided by the pump. The energy equation between the oil surface in the lower tank (1) and the oil surface in the upper tank (2) is

$$\frac{p_1}{\gamma} + \alpha_1 \frac{V_1^2}{2g} + z_1 + h_p = \frac{p_2}{\gamma} + \alpha_2 \frac{V_2^2}{2g} + z_2 + h_t + h_L$$

The pressure at both stations is atmospheric, and the velocities are zero. Also, there is no turbine in the system. So, the energy equation simplifies to

$$h_p = z_2 - z_1 + h_L$$

The head losses are due to pipe friction, check valve, elbow, inlet section, and sudden expansion on entry to the upper tank.

$$h_L = (f\frac{L}{D} + K_V + K_b + K_i + 1)\frac{V^2}{2g}$$

Thus

$$h_p = 10 + (240f + 5 + 0.9 + 0.5 + 1)\frac{V^2}{2g}$$
$$= 10 + (240f + 7.4)\frac{V^2}{2g}$$

The velocity is obtained from

$$V = \frac{Q}{A}$$

$$Q = 100 \text{ gpm} \times \frac{0.00223 \text{ ft}^3/\text{s}}{1 \text{ gpm}} = 0.223 \text{ ft}^3/\text{s}$$

$$A = \left(\frac{2}{12}\right)^2 \frac{\pi}{4} = 0.0218 \text{ ft}^2$$

$$V = \frac{0.223}{0.0218} = 10.23 \text{ ft/s}$$

The relative roughness of the pipe is $0.002/2 = 0.001$. The Reynolds number is

$$\text{Re} = \frac{\rho V D}{\mu} = \frac{0.86 \times 1.94 \times 10.23 \times (2/12)}{2 \times 10^{-4}}$$

$$= 1.42 \times 10^4$$

From the Moody diagram (Fig. 10.8)

$$f = 0.031$$

Thus the head across the pump is

$$h_p = 10 + (240 \times 0.031 + 7.4)\frac{10.23^2}{2 \times 32.2}$$

$$= 34.1 \text{ ft}$$

The power required is

$$P = \frac{\gamma Q h_p}{\eta} = \frac{62.4 \times 0.86 \times 0.223 \times 34.1}{0.8}$$

$$= 510 \text{ ft-lbf/s}$$

$$= \underline{\underline{0.927 \text{ hp}}}$$

# Problem 10.5

A piping system consists of parallel pipes as shown in the following diagram. One pipe has an internal diameter of 0.5 m and is 1000 m long. The other pipe has an internal diameter of 1 m and is 1500 m long. Both pipes are made of cast iron ($k_s = 0.26$ mm). The pipes are transporting water at 20°C ($\rho = 1000$ kg/m$^3$, $\nu = 10^{-6}$ m$^2$/s). The total flow rate is 4 m$^3$/s. Find the flow rate in each pipe and the pressure drop in the system. There is no elevation change. Neglect minor losses.

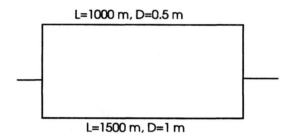

## Solution

Designate the 1000-m pipe as pipe (1) and the other as pipe (2). The pressure drop along each path is the same, so

$$\Delta p = f_1 \frac{L_1}{D_1} \rho \frac{V_1^2}{2} = f_2 \frac{L_2}{D_2} \rho \frac{V_2^2}{2}$$

So, the velocity ratio between the two pipes is

$$\frac{V_2}{V_1} = \sqrt{\frac{f_1}{f_2} \frac{L_1}{L_2} \frac{D_2}{D_1}}$$

$$= \sqrt{\frac{1000}{1500} \times \frac{1.0}{0.5}} \sqrt{\frac{f_1}{f_2}}$$

$$= 1.15 \sqrt{\frac{f_1}{f_2}}$$

Since the total flow rate is 4 m$^3$/s,

$$V_1 A_1 + V_2 A_2 = 4 \text{ m}^3/\text{s}$$

$$V_1 \left(\frac{\pi}{4} \times 0.5^2\right) + V_2 \left(\frac{\pi}{4} \times 1^2\right) = 4$$

$$0.196 V_1 + \frac{\pi}{4} \times 1.15 \times V_1 \sqrt{\frac{f_1}{f_2}} = 4$$

or
$$V_1\left(0.196 + 0.903\sqrt{\frac{f_1}{f_2}}\right) = 4$$

The relative roughness of pipe 1 is 0.26/500=0.00052 and for pipe 2, 26/1000=0.00026. We do not know the friction factors because they depend on the Reynolds number which, in turn, depends on the velocity. An iterative solution is necessary. A good initial guess is to use the friction factor for a fully rough pipe (limit at high Reynolds number). From the Moody diagram (Fig. 10.8) for pipe 1, take $f_1 = 0.017$ and for pipe 2, $f_2 = 0.0145$. Solving for $V_1$

$$V_1 = \frac{4}{0.196 + 0.903\sqrt{\frac{0.017}{0.0145}}}$$
$$= 3.41 \text{ m/s}$$

and

$$V_2 = 1.15 \times 3.41 \times \sqrt{\frac{0.017}{0.0145}}$$
$$= 4.25 \text{ m/s}$$

The Reynolds numbers are

$$\text{Re}_1 = \frac{V_1 D_1}{\nu} = \frac{3.41 \times 0.5}{10^{-6}} = 1.71 \times 10^6$$
$$\text{Re}_2 = \frac{V_2 D_2}{\nu} = \frac{4.25 \times 1.0}{10^{-6}} = 4.25 \times 10^6$$

From the Moody diagram, the corresponding friction factors are

$$f_1 = 0.0172 \qquad f_2 = 0.0145$$

Because these are essentially the same as the initial guesses, further iterations are not necessary. The flow rates in each pipe are

$$Q_1 = A_1 V_1 = 0.196 \times 3.41 = \underline{\underline{0.668 \text{ m}^3/\text{s}}}$$
$$Q_2 = A_2 V_2 = 0.785 \times 4.25 = \underline{\underline{0.334 \text{ m}^3/\text{s}}}$$

The pressure drop is

$$\Delta p = f_1 \frac{L_1}{D_1} \rho \frac{V_1^2}{2}$$
$$= 0.0172 \times \frac{1000}{0.5} \times 1000 \times \frac{3.41^2}{2}$$
$$= 2 \times 10^5 \text{ Pa} = \underline{\underline{200 \text{ kPa}}}$$

# Problem 10.6

Three pipes are connected in series, and the total pressure drop is 200 kPa. The elevation increase between the beginning and end of the system is 10 m. Water at 20°C ($\rho = 1000$ kg/m$^3$, $\nu = 10^{-6}$ m$^2$/s) flows through the system.

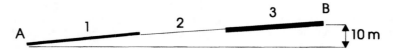

The characteristics of the three pipes are

| Pipe | Length, m | Diameter, m | Roughness, mm | Relative roughness |
|---|---|---|---|---|
| 1 | 100 | 0.1 | 0.25 | 0.0025 |
| 2 | 50 | 0.08 | 0.10 | 0.00125 |
| 3 | 120 | 0.15 | 0.2 | 0.0013 |

Calculate the flow rate. Neglect the transitional losses.

## Solution

Apply the energy equation between the beginning and end of the pipe system ($A$ to $B$).

$$\frac{p_A}{\gamma} + \alpha_A \frac{V_A^2}{2g} + z_A = \frac{p_B}{\gamma} + \alpha_B \frac{V_B^2}{2g} + z_B + h_L$$

Take the flow as turbulent with $\alpha = 1$. Thus

$$\frac{p_A - p_B}{\gamma} + z_A - z_B = \frac{V_B^2 - V_A^2}{2g} + h_L$$

The head loss is the sum of the head loss in each pipe.

$$h_L = f_1 \frac{L_1}{D_1} \frac{V_1^2}{2g} + f_2 \frac{L_2}{D_2} \frac{V_2^2}{2g} + f_3 \frac{L_3}{D_3} \frac{V_3^2}{2g}$$

The velocity in each pipe section is

$$V = \frac{Q}{A}$$

so

$$V_1 = \frac{Q}{\frac{\pi}{4} \times 0.1^2} = 127.4Q$$

$$V_2 = \frac{Q}{\frac{\pi}{4} \times 0.08^2} = 199.0Q$$

$$V_3 = \frac{Q}{\frac{\pi}{4} \times 0.15^2} = 56.6Q$$

The energy equation becomes

$$\frac{200 \times 10^3}{9810} - 10 = \frac{(56.6Q)^2}{2 \times 9.81} - \frac{(127.4Q)^2}{2 \times 9.81} + f_1 \frac{100}{0.1} \frac{(127.4Q)^2}{2 \times 9.81}$$
$$+ f_2 \frac{50}{0.08} \frac{(199Q)^2}{2 \times 9.81} + f_3 \frac{120}{0.15} \frac{(56.6Q)^2}{2 \times 9.81}$$

or

$$10.39 = Q^2(-664 + 8.27 \times 10^5 f_1 + 1.26 \times 10^6 f_2 + 1.31 \times 10^5 f_3)$$

The solution has to be obtained by iteration. First, assume that $f_1 = f_2 = f_3 = 0.02$. Then

$$10.34 = 4.37 \times 10^4 Q^2$$
$$Q = 0.0154 \text{ m}^3/\text{s}$$

Now, calculate the velocity in each pipe, the Reynolds number, and the friction factor.

| Pipe | V (m/s) | Re | f |
|------|---------|-----|-----|
| 1 | 1.96 | $1.96 \times 10^5$ | 0.025 |
| 2 | 3.06 | $2.45 \times 10^5$ | 0.022 |
| 3 | 0.872 | $1.31 \times 10^5$ | 0.023 |

Substituting the values for friction factor back into the equation for $Q$

$$10.39 = 5.07 \times 10^4 Q^2$$
$$Q = 0.0143 \text{ m/s}$$

The new velocities, Reynolds numbers and friction factors are shown in the following table.

| Pipe | V (m/s) | Re | f |
|------|---------|-----|-----|
| 1 | 1.82 | $1.82 \times 10^5$ | 0.0255 |
| 2 | 2.84 | $2.27 \times 10^5$ | 0.022 |
| 3 | 0.81 | $1.21 \times 10^5$ | 0.023 |

The answer is unchanged so

$$Q = \underline{\underline{0.0143 \text{ m}^3/\text{s}}}$$

## Problem 10.7

A duct for an air conditioning system has a rectangular cross-section of 2 ft by 9 in. The duct is fabricated from galvanized iron ($k_s = 0.006$ in.). Calculate the pressure drop for a horizontal 50-ft section of pipe with a flow rate of air of 5000 cfm at 100°F and atmospheric pressure ($\mu = 3.96 \times 10^{-7}$ lbf-s/ft$^2$, $\gamma = 0.0709$ lbf/ft$^3$ and $\nu = 1.8 \times 10^{-4}$ ft$^2$/s).

### Solution

Because the cross-section is not circular, use the hydraulic radius.

$$R_h = \frac{A}{P} = \frac{2 \times 0.75}{4 + 1.5} = 0.273 \text{ ft}$$

The hydraulic diameter is $4R_h = 1.09$ ft. This value is now used as if the pipe had a circular cross-section with this radius. The velocity in the pipe is

$$V = \frac{Q}{A} = \frac{5000/60}{2 \times 0.75} = 55.6 \text{ ft/s}$$

The Reynolds number is

$$\text{Re} = \frac{VD_h}{\nu} = \frac{55.6 \times 1.09}{1.8 \times 10^{-4}} = 3.4 \times 10^5$$

The relative roughness is

$$\frac{k_s}{D_h} = \frac{0.006}{12 \times 1.09} = 0.00046$$

From the Moody diagram (Fig. 10.8) $f = 0.018$. The pressure drop is

$$\Delta p = f \frac{L}{D_h} \rho \frac{V^2}{2}$$
$$= 0.018 \times \frac{50}{1.09} \times \frac{0.0709}{32.2} \times \frac{55.6^2}{2}$$
$$= \underline{\underline{2.81 \text{ psf}}}$$

# Chapter 11

# Drag and Lift

## Problem 11.1

Air with a speed of $V_o$ flows over a long bar that has a $15°$ wedge-shaped cross-section. The pressure variation, as represented using the coefficient of pressure, is shown in the following sketch. On the west face of the bar, the coefficient of pressure is everywhere equal to +1. On the northeast face, the coefficient of pressure varies linearly from -2 to 0, and on the south face the variation is linear from +1 to 0. Determine the coefficient of drag and the coefficient of lift.

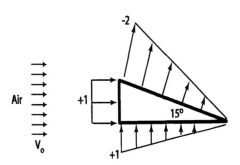

## Solution

As shown in Fig. 1, the pressure distributions cause resultant forces to act on each face of the bar. Identify faces 1, 2, and 3 as the west, northeast, and south sides of the bar, respectively.

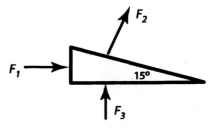

Fig. 1 Resultant force diagram

On the west face of the bar, the average coefficient of pressure is $\overline{C_{p1}} = 1.0$ Thus the force $F_1$ is given by

$$F_1 = \overline{C_{p1}} A_1 \frac{\rho V_o^2}{2}$$
$$= 1.0 A_1 \frac{\rho V_o^2}{2} \quad (1)$$

Similarly, the force on the northeast face is

$$F_2 = \overline{C_{p2}} A_2 \frac{\rho V_o^2}{2}$$
$$= +1.0 A_2 \frac{\rho V_o^2}{2} \quad (2)$$

where $\overline{C_{p2}}$ is given as $+1$ because of the direction of the pressure (outward), as shown in the sketch in the problem statement. Finally,

$$F_3 = \overline{C_{p3}} A_3 \frac{\rho V_o^2}{2}$$
$$= 0.5 A_3 \frac{\rho V_o^2}{2} \quad (3)$$

By definition, lift ($F_L$) and drag ($F_D$) force are perpendicular and parallel, respectively, to the free stream.

Fig. 2 Lift and drag

Equating drag force (Fig. 2) with forces due to pressure (Fig. 1) gives

$$F_D = F_1 + F_2 \sin 15^\circ \quad (4)$$

Substituting Eqs. (1) and (2) into Eq. (4) gives

$$F_D = F_1 + F_2 \sin 15°$$
$$= (1.0 A_1 + 1.0 A_2 \sin 15°) \frac{\rho V_o^2}{2}$$

Letting $A_1/A_2 = \sin 15°$ gives

$$F_D = 2 A_1 \frac{\rho V_o^2}{2} \qquad (5)$$

Since $A_1$ is the projected area, Eq. (5) becomes

$$C_D = \frac{F_D}{\left(\frac{\rho V_o^2}{2}\right) A_1}$$
$$= \underline{\underline{2}}$$

From Figs. 1 and 2

$$F_L = F_2 \cos(15°) + F_3 \qquad (6)$$

Substitute Eqs. (2) and (3) into Eq. (6).

$$F_L = A_2 \frac{\rho V_o^2}{2} \cos(15°) + 0.5 A_3 \frac{\rho V_o^2}{2}$$
$$= A_3 \frac{\rho V_o^2}{2} + 0.5 A_3 \frac{\rho V_o^2}{2}$$
$$= (1.0 + 0.5) A_3 \frac{\rho V_o^2}{2} \qquad (7)$$

or

$$\frac{F_L}{A_3 \frac{\rho V_o^2}{2}} = 1.5 \qquad (8)$$

Use area $A_3$ to define the coefficient of lift.

$$C_L = \frac{F_L}{\left(\frac{\rho V_o^2}{2}\right) A_3} \qquad (9)$$

Combine Eqs. (8) and (9)

$$\underline{\underline{C_L = 1.54}}$$

## Problem 11.2

Air with a speed of 30 m/s and a density of 1.25 kg/m³ flows normal to a rectangular sign of dimension 5.5 m by 7.5 m. Find the force of the air on the sign.

### Solution

The drag force is

$$F_D = C_D A_p \frac{\rho V_o^2}{2}$$

$$= C_D \left(5.5 \times 7.5 \text{ m}^2\right) \frac{\left(1.25 \text{ kg/m}^3\right)\left(30^2 \text{ m}^2/\text{s}^2\right)}{2}$$

$$= C_D \left(23.2 \text{ kN}\right)$$

The coefficient of drag from Table 11.1 with $\ell/b = 7.5/5.5 \approx 1.0$ is 1.18. Thus

$$F_D = 1.18 \left(23.2 \text{ kN}\right)$$
$$= \underline{27.4 \text{ kN}}$$

## Problem 11.3

A student is modeling the drag force on the fins of a model rocket. The rocket has three fins, each fabricated from $\frac{1}{16}$ in.-thick balsa-wood to a dimension of 2.5 × 1 in. The coefficient of drag for each fin is 1.4, and the fin is subjected to air with a speed of 100 mph and a density of 0.00237 slug/ft³. Determine the total drag force on the fins. Since the fins are not streamlined, assume that drag on a given fin is based on the projected area (not the planform area).

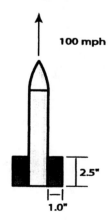

## Solution

The drag force for three fins is

$$F_D = 3 C_D A_p \frac{\rho V_o^2}{2}$$

The projected area (normal to fluid velocity) of one fin is

$$A_p = \left(\frac{1}{16} \times 1\right) \text{ in.}^2 \left(\frac{1 \text{ ft}^2}{144 \text{ in.}^2}\right)$$
$$= 4.34 \times 10^{-4} \text{ ft}^2$$

The velocity is

$$V_o = (100 \text{ mph}) \left(\frac{1.467 \text{ ft/s}}{1 \text{ mph}}\right)$$
$$= 147 \text{ ft/s}$$

Thus the drag force is

$$F_D = 3 C_D A_p \frac{\rho V_o^2}{2}$$
$$= 3 \, (1.4) \left(4.34 \times 10^{-4} \text{ ft}^2\right) \frac{\left(0.00237 \text{ slug/ft}^3\right) \left(147^2 \text{ ft}^2/\text{s}^2\right)}{2}$$
$$= \underline{0.0467 \text{ lbf}}$$

## Problem 11.4

For a bicycle racer who races on the road, a typical speed is 40 kph, the coefficient of drag is about 0.88, and the frontal area is about 0.36 m². Determine the power required to overcome wind drag when there is (a) no headwind and (b) a headwind of 15 kph.

### Solution

Power is the product of drag force and speed of the cyclist

$$P = F_D V_c$$

The speed of the cyclist is

$$V_c = (40 \text{ kph}) \left(\frac{1000 \text{ m}}{1 \text{ km}}\right) \left(\frac{1 \text{ hr}}{3600 \text{ s}}\right)$$
$$= 11.1 \text{ m/s}$$

With no headwind, the drag force is

$$F_D = C_D A_p \frac{\rho V_o^2}{2}$$
$$= 0.88 \, (0.36 \text{ m}^2) \, \frac{\left(1.2 \text{ kg/m}^3\right)\left(11.1^2 \text{ m}^2/\text{s}^2\right)}{2}$$
$$= 23.4 \text{ N}$$

The power is

$$P = F_D V_c$$
$$= (23.4 \text{ N})(11.1 \text{ m/s})$$
$$= \underline{\underline{260 \text{ W}}} \text{ (no headwind)}$$

When there is a headwind, the drag force changes because the velocity term represents the speed of the wind relative to the cyclist. The wind speed is

$$V_{\text{wind}} = (15 \text{ kph}) \left(\frac{1000 \text{ m}}{1 \text{ km}}\right) \left(\frac{1 \text{ hr}}{3600 \text{ s}}\right)$$
$$= 4.17 \text{ m/s}$$

The air speed relative to the cyclist is

$$V_o = V_c + V_{\text{wind}}$$
$$= (11.1 + 4.17) \text{ m/s}$$
$$= 15.3 \text{ m/s}$$

The drag force with the headwind present is

$$F_D = C_D A_p \frac{\rho V_o^2}{2}$$
$$= 0.88 \, (0.36 \text{ m}^2) \, \frac{\left(1.2 \text{ kg/m}^3\right)\left(15.3^2 \text{ m}^2/\text{s}^2\right)}{2}$$
$$= 44.5 \text{ N}$$

The power with the headwind present is

$$P = F_D V_c$$
$$= (44.5 \text{ N})(11.1 \text{ m/s})$$
$$= \underline{\underline{494 \text{ W (with headwind)}}}$$

## Problem 11.5

During the preliminary design of a submarine, a designer assumes that the drag force will be equal to the drag on a streamlined body that has a diameter of 1.5 m and a length of 8 m. The design speed is 10 m/s, the submarine will operate in 10 °C water (kinematic viscosity is $\nu = 1.31 \times 10^{-6}$ m$^2$/s), and the sub will be powered by an electric motor with an efficiency of 90%. Determine the power that will be consumed by the motor.

### Solution

Power is the product of drag force and speed of the submarine

$$P = F_D V_s$$

The power that will be consumed by the electric motor is increased because of the efficiency rating ($\eta$).

$$P = \frac{F_D V_s}{\eta}$$

To find drag force, the Reynolds number is needed.

$$\text{Re} = \frac{V_s D}{\nu}$$
$$= \frac{(10 \text{ m/s})(1.5 \text{ m})}{(1.31 \times 10^{-6} \text{ m}^2/\text{s})}$$
$$= 11.5 \times 10^6$$

Fig. 11.11 hows the coefficient of drag for a streamlined body with $L/d = 5$. Since the aspect ratio of the submarine is $L/d = 8/1.5 = 5.33$, Fig. 11.11 provides a good approximation. Also, we need to extrapolate the Reynolds number (the data goes to Re = $10^7$). Estimating from Figure 11.11,

$$C_D \approx 0.045$$

The drag force is

$$F_D = C_D A_p \frac{\rho V_o^2}{2}$$
$$= 0.045 \left(\frac{\pi \times 1.5^2}{4} \text{ m}^2\right) \frac{\left(1000 \text{ kg/m}^3\right)\left(10^2 \text{ m}^2/\text{s}^2\right)}{2}$$
$$= 3980 \text{ N}$$

The power is

$$P = \frac{F_D V_s}{\eta}$$
$$= \frac{(3980 \text{ N})(10 \text{ m/s})}{0.9}$$
$$= \underline{44.2 \text{ kW}} \quad (59.3 \text{ hp})$$

# Problem 11.6

Find the terminal velocity of a 18-cm diameter, helium-filled balloon. The balloon material has a mass of 2 g, the helium in the balloon is at a pressure of 2.5 kPa, and the balloon is moving through air at 20 °C.

## Solution

The free-body diagram is

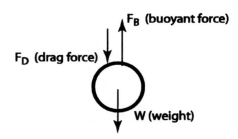

At terminal velocity, the forces sum to zero.

$$F_B = F_D + W \tag{1}$$

The buoyant force is

$$F_B = \gamma_{\text{air}} \frac{4\pi r^3}{3}$$

$$= \left(11.8 \text{ N/m}^3\right) \left(\frac{4\pi \left(0.09^3 \text{ m}^3\right)}{3}\right)$$

$$= 0.0360 \text{ N}$$

Since the given pressure force is small $(p \ll p_{atm})$, assume it is a gage pressure. The absolute pressure of the helium is

$$p = (2.5 + 101.3) \text{ kPa}$$
$$= 102.8 \text{ kPa}$$

Density of the helium is

$$\rho = \frac{p}{RT}$$

$$= \frac{(102.8)\left(10^3 \text{ Pa}\right)}{(2077 \text{ J/kg-K})(20 + 273.2 \text{ K})}$$

$$= 0.17 \text{ kg/m}^3$$

The total weight is the sum of the weight of the helium and the weight of the balloon.

$$W = \gamma_{\text{helium}} \frac{4\pi r^3}{3} + mg$$

$$= \left(0.17 \text{ kg/m}^3\right)\left(9.81 \text{ m/s}^2\right)\left(\frac{4\pi \left(0.09^3 \text{ m}^3\right)}{3}\right) + (0.002 \text{ g})\left(9.81 \text{ m/s}^2\right)$$

$$= 0.00509 \text{ N} + 0.0196 \text{ N}$$

$$= 0.0247 \text{ N}$$

Eq. (1) becomes

$$F_B = F_D + W$$
$$0.036 = F_D + 0.0247$$

The drag force is

$$F_D = C_D \left(\pi r^2\right) \left(\frac{\rho V_o^2}{2}\right)$$

Combining equations gives

$$0.036 = C_D \left(\pi r^2\right) \left(\frac{\rho V_o^2}{2}\right) + 0.0247$$

Substituting values gives

$$0.036 = C_D \left(\pi \times 0.09^2\right) \left(\frac{1.2 V_o^2}{2}\right) + 0.0247$$

$$0.036 = 0.0153 C_D V_o^2 + 0.0247 \qquad (2)$$

Rearranging Eq. (2) gives
$$0.739 = C_D V_o^2 \tag{3}$$

While Eq. (3) has two unknowns ($V_o$ and $C_D$), it has a unique solution (i.e. it is solvable) because $C_D$ is a function of $V_o$. One method to solve Eq. (3) is to use an iterative approach, as described by a four-step process:

1. Guess a value of $V_o$.
2. Calculate $Re_D$ and then find $C_D$.
3. Solve Eq. (3) for $V_o$.
4. If the $V_o$ values from steps 1 and 3 agree, then stop; otherwise, go back to step 1.

The solution approach is implemented as follows.

1. Guess that $V_o = 2$ m/s.
2. $Re_D = (2 \text{ m/s}) (0.18 \text{ m}) / (15.1 \times 10^{-6} \text{ m}^2/\text{s}) = 23{,}800$; From Fig. 11.11, $C_D \approx 0.42$.
3. From Eq. (3), $V_o = \sqrt{0.739/0.42} = 1.33$ m/s.
4. Since $V_o$ from steps 1 and 3 disagree, guess $V_o = 1.33$ m/s.
5. $Re_D = (1.33 \text{ m/s}) (0.18 \text{ m}) / (15.1 \times 10^{-6} \text{ m}^2/\text{s}) = 15{,}900$; From Fig. 11.11, $C_D \approx 0.41$.
6. From Eq. (3), $V_o = \sqrt{0.739/0.41} = 1.34$ m/s.
7. Since $V_o$ from steps 1 and 3 agree, we can stop.

$$V_{\text{terminal}} = \underline{\underline{1.34 \text{ m/s}}}$$

# Chapter 12

# Compressible Flow

## Problem 12.1

Methane at 25°C ($R = 518$ J/kg/K, $k = 1.31$) is flowing in a pipe at 400 m/s. Is the flow subsonic, sonic, supersonic, or hypersonic?

### Solution

The speed of sound in methane is

$$c = \sqrt{kRT} = \sqrt{1.31 \times 518 \times 298}$$
$$= \underline{\underline{450 \text{ m/s}}}$$

Because the velocity is less than the speed of sound, the flow is <u>subsonic</u>.

## Problem 12.2

Air ($R = 1716$ ft-lbf/slug/°R, $k = 1.40$) with a velocity of 1500 ft/s, a pressure of 10 psia, and a temperature of 40°F passes through a normal shock wave. Find the velocity, pressure, and temperature downstream of the shock wave.

### Solution

First find the upstream Mach number and then use relationships for normal shock waves. The speed of sound is

$$c = \sqrt{kRT} = \sqrt{1.4 \times 1716 \times (460 + 40)}$$
$$= 1096 \text{ ft/s}$$

The upstream Mach number is

$$M_1 = \frac{V}{c} = \frac{1500}{1096} = 1.37$$

The Mach number behind the shock wave is

$$M_2^2 = \frac{(k-1)M_1^2 + 2}{2kM_1^2 - (k-1)}$$

$$= \frac{0.4 \times 1.37^2 + 2}{2 \times 1.4 \times 1.37^2 - 0.4}$$

$$= 0.567$$

$$M_2 = 0.753$$

The temperature ratio across the wave is

$$\frac{T_2}{T_1} = \frac{1 + \frac{k-1}{2}M_1^2}{1 + \frac{k-1}{2}M_2^2}$$

$$= \frac{1 + 0.2 \times 1.37^2}{1 + 0.2 \times 0.753^2}$$

$$= 1.24$$

Thus the temperature is

$$T_2 = 1.24 T_1$$
$$= 1.24 \times 500$$
$$= 620°R = \underline{\underline{160°F}}$$

The pressure ratio is

$$\frac{p_2}{p_1} = \frac{1 + kM_1^2}{1 + kM_2^2}$$

$$= \frac{1 + 1.4 \times 1.37^2}{1 + 1.4 \times 0.753^2}$$

$$= 2.022$$

Thus the pressure is

$$p_2 = 2.022 p_1 = 2.022 \times 10$$
$$= \underline{\underline{20.22 \text{ psia}}}$$

The speed of sound behind the shock wave is

$$c_2 = \sqrt{1.4 \times 1716 \times 620} = 1220 \text{ ft/s}$$

The velocity behind the shock wave is

$$V_2 = M_2 c_2$$
$$= 0.753 \times 1220$$
$$= \underline{\underline{919 \text{ ft/s}}}$$

## Problem 12.3

Air ($R = 287$ J/kg/K, $k = 1.4$) at 800 kPa and 20°C exhausts through a truncated nozzle with an area of 0.6 cm² to a back pressure of 100 kPa. Calculate the flow rate.

## Solution

First find out if the exit condition is sonic or subsonic. The exit pressure for a sonic nozzle would be

$$\frac{p_o}{p_*} = \left(\frac{k+1}{2}\right)^{\frac{k}{k-1}} = 1.2^{3.5} = 1.89$$

so

$$p_* = \frac{800}{1.89} = 423 \text{ kPa}$$

Since the exit pressure is larger than the back pressure, the flow at the exit will be sonic.

The flow rate is

$$\dot{m} = \rho_* A c_*$$

where the conditions are evaluated at the exit (sonic condition). The exit temperature is found from

$$\frac{T_o}{T_*} = \frac{k+1}{2} = 1.2$$

Thus the exit temperature is

$$T_* = \frac{273 + 20}{1.2} = 244 \text{ K}$$

The sonic velocity at this temperature is

$$c_* = \sqrt{1.4 \times 287 \times 244} = 313 \text{ m/s}$$

The exit density is

$$\rho_* = \frac{p_*}{RT_*} = \frac{423 \times 10^3}{287 \times 244} = 6.04 \text{ kg/m}^3$$

The flow rate is

$$\dot{m} = 6.04 \times 0.6 \times 10^{-4} \times 313$$
$$= \underline{\underline{0.113 \text{ kg/s}}}$$

## Problem 12.4

A rocket nozzle is designed to expand exhaust gases ($R = 300$ J/kg/K, $k = 1.3$) from a chamber pressure of 600 kPa and total temperature of 3000 K to a Mach number of 2.5 at the exit. The throat area is 0.1 m$^2$. Find the area at the exit, the exit pressure, the exit velocity, and the mass flow rate.

### Solution

The relationship for the ratio of the nozzle area to the throat area is

$$\frac{A}{A_*} = \frac{1}{M}\left(\frac{1+\frac{k-1}{2}M^2}{\frac{k+1}{2}}\right)^{\frac{k+1}{2(k-1)}}$$

For a Mach number of 2.5

$$\frac{A}{A_*} = \frac{1}{2.5}\left(\frac{1+0.15\times 2.5^2}{1.15}\right)^{3.83}$$
$$= 2.95$$

Thus the exit area is
$$A_e = 2.95 \times 0.1 = \underline{0.295 \text{ m}^2}$$

The exit pressure is obtained from

$$\frac{p_o}{p} = \left(1 + \frac{k-1}{2}M^2\right)^{\frac{k}{k-1}}$$
$$= (1 + 0.15 \times 2.5^2)^{4.33}$$
$$= 17.5$$

The exit pressure is
$$p_e = \frac{600}{17.5} = \underline{34.3 \text{ kPa}}$$

The exit temperature is obtained from
$$\frac{T_o}{T} = 1 + \frac{k-1}{2}M^2$$
$$= 1 + 0.15 \times 2.5^2$$
$$= 1.94$$

The exit temperature is
$$T_e = \frac{3000}{1.94} = \underline{1546 \text{ K}}$$

The speed of sound at the exit is
$$c_e = \sqrt{kRT_e} = \sqrt{1.3 \times 300 \times 1546}$$
$$= 776 \text{ m/s}$$

so the exit velocity is
$$V_e = M_e c_e = 2.5 \times 776 = \underline{\underline{1940 \text{ m/s}}}$$

The density at the exit is
$$\rho_e = \frac{p_e}{RT_e} = \frac{34.3 \times 10^3}{300 \times 1546} = 0.074 \text{ kg/m}^3$$

The mass flow is
$$\dot{m} = \rho_e V_e A_e = 0.074 \times 1940 \times 0.295$$
$$= \underline{\underline{42.3 \text{ kg/s}}}$$

## Problem 12.5

An airplane is flying through air ($R = 1716$ ft-lbf/slug/°R, $k = 1.4$) at 600 ft/s. The pressure and temperature of the air are 14 psia and 50°F. What are the pressure and temperature at the stagnation point? (Assume the stagnation process is isentropic.)

## Solution

The stagnation pressure corresponds to the total conditions if the process is isentropic. The speed of sound in air at this condition is
$$c = \sqrt{kRT} = \sqrt{1.4 \times 1716 \times (460 + 50)}$$
$$= 1107 \text{ ft/s}$$

The Mach number is
$$M = \frac{V}{c} = \frac{600}{1107} = 0.542$$

The total temperature is
$$T_o = T(1 + \frac{k-1}{2} M^2)$$
$$= 510 \times (1 + 0.2 \times 0.542^2)$$
$$= 540°\text{R} = \underline{\underline{80°\text{F}}}$$

The total pressure is
$$p_o = p\left(1 + \frac{k-1}{2} M^2\right)^{\frac{k}{k-1}}$$
$$= 14 \times (1 + 0.2 \times 0.542^2)^{3.5}$$
$$= \underline{\underline{17.1 \text{ psia}}}$$

## Problem 12.6

Carbon dioxide ($R = 189$ J/kg/K, $k = 1.3$) flows through a Laval nozzle. A normal shock wave occurs in the expansion section where the Mach number is 2 and the cross-sectional area is 1 cm². The exit area is 1.5 cm². The total pressure before the shock wave is 400 kPa. Find the area at the throat and the exit pressure.

### Solution

The ratio of the nozzle area to the throat area is

$$\frac{A}{A_*} = \frac{1}{M}\left(\frac{1 + \frac{k-1}{2}M^2}{\frac{k+1}{2}}\right)^{\frac{k+1}{2(k-1)}}$$

$$= \frac{1}{2}\left(\frac{1 + 0.15 \times 2^2}{1.15}\right)^{3.83}$$

$$= 1.77$$

The throat area is

$$A_* = \frac{1}{1.77} = 0.56 \text{ cm}^2$$

After the normal shock wave, the flow will be subsonic. The Mach number just downstream of the normal shock is

$$M_2^2 = \frac{(k-1)M_1^2 + 2}{2kM_1^2 - (k-1)}$$

$$= \frac{0.3 \times 2^2 + 2}{2 \times 1.3 \times 2^2 - 0.3}$$

$$= 0.317$$

$$M_2 = 0.563$$

The change in pressure across the shock wave is

$$\frac{p_2}{p_1} = \frac{1 + kM_1^2}{1 + kM_2^2}$$

$$= \frac{1 + 1.3 \times 2^2}{1 + 1.3 \times 0.563^2}$$

$$= 4.39$$

The change in total pressure across the normal shock wave is

$$\frac{p_{o,2}}{p_{o,1}} = \frac{p_2}{p_1}\left(\frac{1 + \frac{k-1}{2}M_2^2}{1 + \frac{k-1}{2}M_1^2}\right)^{\frac{k}{k-1}}$$

$$= 4.39 \times \left(\frac{1 + 0.15 \times 0.563^2}{1 + 0.15 \times 2^2}\right)^{4.33}$$

$$= 0.701$$

The total pressure for the subsonic flow behind the shock wave is

$$p_{o,2} = 400 \times 0.701 = 280 \text{ kPa}$$

The subsonic flow in the expansion section will now decrease in Mach number as the area increases. The Mach number at the exit must be obtained from the relationship between area-ratio and Mach number that was used previously. First, we have to find the area where the flow would become sonic for the conditions downstream of the normal shock wave.

$$\frac{A}{A_*} = \frac{1}{M} \left( \frac{1 + \frac{k-1}{2} M^2}{\frac{k+1}{2}} \right)^{\frac{k+1}{2(k-1)}}$$

$$= \frac{1}{0.563} \left( \frac{1 + 0.15 \times 0.563^2}{1.15} \right)^{3.83}$$

$$= 1.24$$

so the area where sonic flow would occur is

$$A_* = \frac{1}{1.24} = 0.806 \text{ cm}^2$$

and the area ratio of the exit is then

$$\frac{A_e}{A_*} = \frac{1.5}{0.806} = 1.86$$

We now have to find the subsonic Mach number that corresponds to this area ratio.

$$1.86 = \frac{1}{M} \left( \frac{1 + 0.15 M^2}{1.15} \right)^{3.83}$$

This equation must be solved iteratively. The result is 0.336. The exit pressure is then

$$\frac{p_{o,2}}{p_e} = (1 + 0.15 \times 0.336^2)^{4.33}$$

$$= 1.075$$

So the exit pressure is

$$p_e = \frac{280}{1.075} = \underline{260 \text{ kPa}}$$

# Chapter 13

# Flow Measurements

## Problem 13.1

Velocity in an air flow is to be measured with a stagnation tube that has a resolution of 0.1-in. $H_2O$. Find the minimum fluid speed in ft/s that can be measured. Neglect viscous effects and assume that the air is at room condition.

### Solution

Fluid speed for a stagnation tube is given by Eq. (5.19) in the 7$^{th}$ edition.

$$V = \sqrt{\frac{2\Delta p}{\rho}}$$

Convert the pressure change from a unit of in. $H_2O$ to a unit of psf.

$$\Delta p = (0.1 \text{ in. } H_2O)\left(\frac{0.03609 \text{ psi}}{\text{in. } H_2O}\right)\left(\frac{144 \text{ psf}}{\text{psi}}\right)$$
$$= 0.520 \text{ psf}$$

Note that this pressure value could also have been found using the hydrostatic equation: $\Delta p = (\rho_{H_2O})\, gh$.
The minimum velocity is

$$V = \sqrt{\frac{2\Delta p}{\rho_{\text{air}}}}$$
$$= \sqrt{2\left(\frac{0.520 \text{ lbf}}{\text{ft}^2}\right)\left(\frac{\text{ft}^3}{0.00233 \text{ slug}}\right)\left(\frac{\text{slug} \cdot \text{ft}}{\text{lbf} \cdot \text{s}^2}\right)}$$
$$= \underline{\underline{21.1 \text{ ft/s}}}$$

## Problem 13.2

Air velocity is measured with a stagnation tube of diameter $d = 0.5$ mm. Pressure in the stagnation tube causes water in a U-tube to rise to a height $h$. Find the minimum velocity $V$ that can be measured with the stagnation tube if the aim is that viscous effects contribute an error less than 5%. Also, find the corresponding value of $\Delta p$.

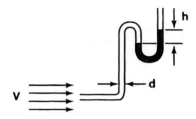

## Solution

Viscous effects are characterized in Fig. 13.1. From the vertical axis of this figure

$$V_{\text{actual}} = \sqrt{\frac{2\Delta p}{\rho C_p}}$$

When neglecting viscous effects, the corresponding formula is

$$V_{\text{approx}} = \sqrt{\frac{2\Delta p}{\rho}}$$

The error $e$ is given by

$$e = \frac{V_{\text{approx}} - V_{\text{actual}}}{V_{\text{actual}}}$$
$$= \frac{1 - \sqrt{1/C_p}}{\sqrt{1/C_p}} \tag{1}$$

Algebraic manipulation of Eq. (1) gives

$$C_p = (1 + e)^2$$

So, a 5% error is associated with

$$C_p = (1 + 0.05)^2$$
$$= 1.103$$

From Fig. 13.1 in the textbook, this occurs at a Reynolds number of about 25. Thus

$$\text{Re} = 25 = \frac{Vd}{\nu}$$

So

$$V_{\min} = \frac{25\nu}{d}$$
$$= \frac{25 \times (15.1 \times 10^{-6})}{0.5 \times 10^{-3}}$$
$$= \underline{\underline{0.775 \text{ m/s}}}$$

Pressure change is related to fluid speed by

$$V = \sqrt{\frac{2\Delta p}{\rho}}$$

So

$$\Delta p = \frac{\rho V^2}{2}$$
$$= \frac{(1.2 \text{ kg/m}^3)(0.775^2 \text{ m}^2/\text{s}^2)}{2}$$
$$= \underline{\underline{0.360 \text{ Pa}}}$$

## Problem 13.3

The average velocity of gasoline (S = 0.68, $\nu = 4.6 \times 10^{-6}$ ft$^2$/s) is measured with a 2-in. diameter orifice meter in a 6-in. diameter pipe. The manometer uses mercury with dimensions of $h = 4$ in. and $s = 3$ in. Find $V_1$.

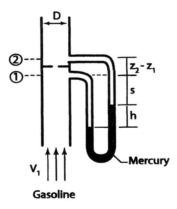

## Solution

Discharge and velocity are related by

$$Q = A_1 V_1 \tag{1}$$

and discharge for an orifice meter is given by

$$Q = K A_o \sqrt{2g\Delta h} \tag{2}$$

Before $K$ can be looked up, piezometric head ($\Delta h$) is needed. This is defined by

$$\Delta h = \left(\frac{p}{\gamma_{\text{gasoline}}} + z\right)_1 - \left(\frac{p}{\gamma_{\text{gasoline}}} + z\right)_2 \tag{3}$$

Applying the manometer equation (Eq. 3.17 in 8th edition) yields

$$p_1 + \gamma_{\text{gasoline}}(s+h) - \gamma_{\text{Hg}}(h) - \gamma_{\text{gasoline}}(s+(z_2-z_1)) = p_2 \tag{4}$$

Rearranging Eq. (4)

$$\left(\frac{p}{\gamma_{\text{gasoline}}} + z\right)_1 - \left(\frac{p}{\gamma_{\text{gasoline}}} + z\right)_2 = h\left(\frac{\gamma_{\text{Hg}}}{\gamma_{\text{gasoline}}} - 1\right) \tag{5}$$

Combining Eqs. (3) and (5)

$$\Delta h = h\left(\frac{\gamma_{\text{Hg}}}{\gamma_{\text{gasoline}}} - 1\right)$$
$$= (4/12 \text{ ft})\left(\frac{13.55}{0.68} - 1\right)$$
$$= 6.31 \text{ ft}$$

To find the flow coefficient $K$, calculate the parameter on the top axis of Fig. 13.13.

$$\frac{\text{Re}_d}{K} = \sqrt{2g\Delta h}\frac{d}{\nu}$$
$$= \sqrt{2 \times 32.2 \times 6.31}\frac{2/12}{4.6 \times 10^{-6}}$$
$$= 730{,}000$$

On Fig. 13.3, tracing the dashed line to $d/D = 2/6 = 0.333$ and interpolating gives $K \approx 0.606$.

Combining Eqs. (1) and (2) and substituting values gives

$$V_1 = K\frac{A_o}{A_1}\sqrt{2g\Delta h}$$
$$= 0.606\frac{(2^2 \text{ in.}^2)}{(6^2 \text{ in.}^2)}\sqrt{2 \times (32.2 \text{ ft/s}^2) \times (6.31 \text{ ft})}$$
$$= \underline{\underline{1.36 \text{ ft/s}}}$$

# Problem 13.4

Water speed is measured with a venturi meter. Throat diameter is 6 cm, pipe diameter is 12 cm, and height on the manometer is $a = 100$ cm. Find the flow rate in the pipe. Kinematic viscosity of water is $\nu = 10^{-6}$ m$^2$/s.

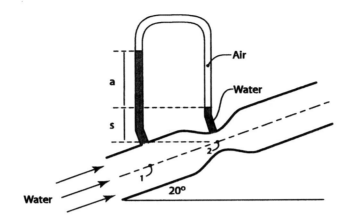

## Solution

Flow rate through a venturi meter is given by

$$Q = K A_o \sqrt{2g \Delta h} \tag{1}$$

Before $K$ can be looked up, piezometric head ($\Delta h$) is needed. This is defined by

$$\Delta h = \left( \frac{p}{\gamma_{H_2O}} + z \right)_1 - \left( \frac{p}{\gamma_{H_2O}} + z \right)_2 \tag{2}$$

where locations 1 and 2 are defined in the sketch. Applying the manometer equation (Eq. 3.17 in 8th edition) yields

$$p_1 - \gamma_{H_2O}(s + z_2 - z_1) - \gamma_{H_2O}(a) + \gamma_{\text{air}}(a) + \gamma_{H_2O}(s) = p_2 \tag{3}$$

Rearranging Eq. (3) gives

$$\left( \frac{p}{\gamma_{H_2O}} + z \right)_1 - \left( \frac{p}{\gamma_{H_2O}} + z \right)_2 = a \left( 1 - \frac{\gamma_{\text{air}}}{\gamma_{H_2O}} \right) \tag{4}$$

Combining Eqs. (2) and (4), and letting $\gamma_{\text{air}}/\gamma_{H_2O} \approx 0$ gives

$$\Delta h = a$$
$$= 1 \text{ m}$$

To find the flow coefficient $K$, calculate the parameter on the top axis of Fig. 13.13.

$$\frac{\text{Re}_d}{K} = \sqrt{2g\Delta h}\frac{d}{\nu}$$
$$= \sqrt{2 \times 9.8 \times 1}\frac{0.06}{10^{-6}}$$
$$= 266,000$$

On Fig. 13.3, tracing the dashed line to $d/D = 6/12 = 0.5$ and interpolating gives $K \approx 1.01$.

Substituting values into Eq. (1) gives

$$Q = KA_o\sqrt{2g\Delta h}$$
$$= 1.01\frac{\pi\left(0.06^2 \text{ m}^2\right)}{4}\sqrt{2 \times \left(9.8 \text{ m/s}^2\right) \times (1.0 \text{ m})}$$
$$= \underline{\underline{0.0126 \text{ m}^3/\text{s}}}$$

# Problem 13.5

Air of density $\rho = 1.2$ m$^3$/s and speed $V_1 = 20$ m/s is metered with an orifice. The orifice diameter is 2 cm, and the pipe diameter is 4 cm. A differential pressure gage records the pressure difference between pressure taps 1 and 2, which are separated by a vertical distance of $a = 8$ cm. Find the reading on the pressure gage. The kinematic viscosity of air is $14.6 \times 10^{-6}$ m$^2$/s.

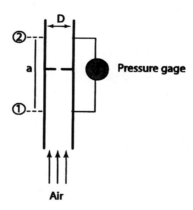

# Solution

When pressure difference is measured with a transducer as shown, the pressure reading is piezometric pressure, and flow rate through the orifice meter is

$$Q = KA_o\sqrt{\frac{2\Delta p}{\rho}} \qquad (1)$$

Rearranging Eq. (1) gives the pressure change.

$$\Delta p = \frac{\rho}{2}\left(\frac{Q}{KA_o}\right)^2 \qquad (2)$$

The flow rate is

$$\begin{aligned}Q &= A_1 V_1 \\ &= \left(\frac{\pi \times 0.04^2}{4}\ \text{m}^2\right)(20\ \text{m/s}) \\ &= 0.0251\ \text{m}^3/\text{s}\end{aligned}$$

To find the flow coefficient $K$, calculate the Reynolds number as defined on the bottom axis of Fig. 13.13.

$$\begin{aligned}\text{Re}_d &= \frac{4Q}{\pi d \nu} \\ &= \frac{4\,(0.0251\ \text{m}^3/\text{s})}{\pi\,(0.02\ \text{m})\,(14.6 \times 10^{-6}\ \text{m}^2/\text{s})} \\ &= 109{,}000\end{aligned}$$

Interpolating in Fig. 13.3 with $d/D = 2/4 = 0.5$ gives $K \approx 0.63$.
The area of the orifice is

$$\begin{aligned}A_o &= \left(\frac{\pi \times 0.02^2}{4}\ \text{m}^2\right) \\ &= 3.142 \times 10^{-4}\ \text{m}^2\end{aligned}$$

Substituting values into Eq. (2)

$$\begin{aligned}\Delta p &= \frac{\rho}{2}\left(\frac{Q}{KA_o}\right)^2 \\ &= \frac{1.2\ \text{kg/m}^3}{2}\left(\frac{0.0251\ \text{m}^3/\text{s}}{(0.63)\times(3.141\times 10^{-4}\ \text{m}^2)}\right)^2 \\ &= 9.65 \times 10^3\ \text{Pa}\end{aligned}$$

So

$$\underline{\underline{\Delta p = 9.65\ \text{kPa}}}$$

# Problem 13.6

An engineer is considering the feasibility of a small hydroelectric power plant, and she wishes to design a rectangular weir to measure the discharge of a small creek. The weir will span the creek, which is 1.5 m wide, and the engineer estimates that the maximum discharge will be 0.5 m³/s. If the creek level cannot rise above 1.2 m, calculate the height of the weir.

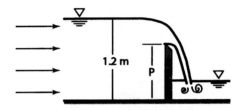

## Solution

Discharge is

$$Q = K\sqrt{2g}LH^{3/2} \qquad (1)$$

where the head on the weir $H$ is given by

$$H = 1.2 - P \qquad (2)$$

and the flow coefficient is

$$K = 0.40 + 0.05\frac{H}{P} \qquad (3)$$

Combining Eqs. (1) to (3)

$$Q = \left(0.40 + 0.05\frac{1.2 - P}{P}\right)\sqrt{2g}L\,(1.2 - P)^{3/2}$$

$$0.5 = \left(0.40 + 0.05\frac{1.2 - P}{P}\right)\sqrt{2 \times 9.81}\,(1.2)\,(1.2 - P)^{3/2} \qquad (4)$$

One way to solve Eq. (4) is to program the right side of the equation and then substitute values into the equation, until a value of 0.5 is achieved. This was done–the results are

$$K = 0.422$$
$$H = 0.368$$
$$P = \underline{0.832 \text{ m}}$$

# Chapter 14

# Turbomachinery

## Problem 14.1

A propeller is to be selected for a light airplane with a mass of 1500 kg which will cruise at 100 m/s at an altitude where the density is 1 kg/m³ The lift-to-drag ratio at cruise conditions is 30:1, and the engine rpm is 3500. The thrust coefficient at maximum efficiency is 0.03, and the maximum efficiency is 60%. Find the diameter of the propeller, the advance ratio at maximum efficiency, and the power output required by the engine.

### Solution

At cruise conditions, the drag is equal to the thrust.

$$T = D = L/(L/D) = 1500 \times 9.81/30 = 490.5 \text{ N}$$

The thrust is given by

$$T = C_T \rho n^2 D^4$$

$$490.5 = 0.03 \times 1 \times \left(\frac{3500}{60}\right)^2 D^4$$

$$= 102.1 D^4$$

Solving for diameter

$$D^4 = 4.805$$
$$D = \underline{1.48 \text{ m}}$$

The advance ratio is

$$J = \frac{V_o}{nD} = \frac{100}{58.3 \times 1.48}$$
$$= \underline{1.16}$$

The power output is

$$P = \frac{TV_o}{\eta} = \frac{490.5 \times 100}{0.6}$$
$$= 81.75 \text{ kW}$$
$$= \underline{\underline{110 \text{ hp}}}$$

## Problem 14.2

A pump delivers 0.25 m³/s of water against a head of 250 m at a rotational speed of 2000 rpm. Find the specific speed, and recommend the appropriate type of pump.

### Solution

The specific speed of the pump is

$$n_s = \frac{nQ^{1/2}}{(g\Delta H)^{3/4}}$$
$$= \frac{\frac{2000}{60} \times 0.25^{1/2}}{(9.81 \times 250)^{3/4}}$$
$$= 0.048$$

From Fig. 14.14, a mixed flow pump is recommended.

## Problem 14.3

A Francis turbine is being designed for a hydroelectric power system. The flow rate of water into the turbine is 5 m³/s. The outer radius of the blade is 0.8 m, and the inner radius is 0.5 m. The width of the blade is 15 cm. The inlet vane angle is 80°. The turbine rotates at 10 rps. Find the inlet angle of the flow with respect to the turbine to ensure nonseparating flow, the outlet vane angle to maximize the power, and the power delivered by the turbine.

### Solution

The radial component of velocity into the turbine is

$$V_{r,1} = \frac{Q}{A_1} = \frac{Q}{2\pi r_1 B}$$
$$= \frac{5}{2\pi \times 0.8 \times 0.15}$$
$$= 6.63 \text{ m/s}$$

The rotational speed is $10 \times 2\pi = 62.8$ rad/s. The angle for nonseparating flow is

$$\alpha_1 = \text{arccot}(\frac{r_1 \omega}{V_{r,1}} + \cot \beta_1)$$
$$= \text{arccot}(\frac{0.8 \times 62.8}{6.63} + \cot 80°)$$
$$= \underline{7.35°}$$

The power produced by the turbine is

$$P = \rho Q \omega \left( r_1 V_1 \cos \alpha_1 - r_2 V_2 \cos \alpha_2 \right)$$

At maximum power the outlet angle, $\alpha_2$, should be $\pi/2$. In other words, the flow would exit radially inward. The outlet angle for the exit is

$$\alpha_2 = \text{arccot}(\frac{r_2 \omega}{V_{r,2}} + \cot \beta_2)$$

so

$$\cot \alpha_2 = \cot \frac{\pi}{2} = 0 = \frac{r_2 \omega}{V_{r,2}} + \cot \beta_2$$

The radial velocity for the inner radius is

$$V_{r,2} = \frac{Q}{A_2} = \frac{Q}{2\pi r_2 B}$$
$$= \frac{5}{2\pi \times 0.5 \times 0.15} = 10.6 \text{ m/s}$$

Thus

$$\cot \beta_2 = -\frac{0.5 \times 62.8}{10.6} = -2.96$$
$$\beta_2 = \underline{161°}$$

The inlet velocity is

$$V_1 = \frac{V_{r,1}}{\sin \alpha_1} = \frac{6.63}{\sin 7.35°} = 51.8 \text{ m/s}$$

The power output is

$$P = \rho Q \omega r_1 V_1 \cos \alpha_1$$
$$= 1000 \times 5 \times 62.8 \times 0.8 \times 51.8 \times \cos 7.35°$$
$$= \underline{1.29 \text{ MW}}$$

# Problem 14.4

A pump is being used to pump water at 80°F ($p_v = 0.506$ psia) from a supply reservoir at 20 psia. The inlet to the pump is a 3-inch pipe. The NPSH for the pump is 10 ft. Find the maximum flow rate in gpm to avoid cavitation. Neglect head losses associated with the inlet and supply pipe.

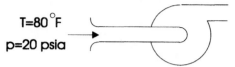

## Solution

The net positive suction head (NPSH) is defined as the difference between the local head at the entrance to the pump and the vapor pressure.

$$\text{NPSH} = \frac{p - p_v}{\gamma}$$

The energy equation between the supply reservoir (1) and the entrance to the pump (2) is

$$\frac{p_1}{\gamma} + \frac{V_1^2}{2g} + z_1 = \frac{p_2}{\gamma} + \frac{V_2^2}{2g} + z_2 + h_L$$

Simplifying

$$\frac{p_2}{\gamma} = \frac{p_1}{\gamma} - \frac{V_2^2}{2g}$$

The head at the pump entrance must be

$$\frac{p_2}{\gamma} = \text{NPSH} + \frac{p_v}{\gamma}$$
$$= 10 + \frac{0.506 \times 144}{62.4}$$
$$= 11.17 \text{ ft}$$

The velocity head must be

$$\frac{V_2^2}{2g} = \frac{20 \times 144}{62.4} - 11.17$$
$$= 35.0 \text{ ft}$$

So the velocity is

$$V_2 = \sqrt{2 \times 32.2 \times 35} = 47.5 \text{ ft/s}$$

The corresponding flow rate is

$$Q = VA = 47.5 \times \frac{\pi}{4} \times \left(\frac{3}{12}\right)^2$$
$$= 2.33 \text{ cfs} = 140 \text{ cfm}$$
$$= \underline{1047 \text{ gpm}}$$

# Problem 14.5

A wind tunnel is being designed as shown. The air is drawn in through a series of screens and flow straighteners at a diameter of 1.5 m. The test section of the tunnel is 1 m. A fan is mounted downstream of the test section. The head loss coefficient for the screens and straighteners is 0.2 and the head loss coefficient for the rest of the tunnel is 0.05 based on the velocity in the test section. The axial fan has a pressure-flow rate curve represented by

$$\Delta p = 1000 \left[1 - \left(\frac{Q}{100}\right)^2\right] \text{ Pa}$$

where $Q$ is in m$^3$/s. Find the velocity in the test section. Take $\rho = 1.2$ kg/m$^3$.

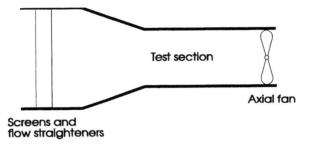

Screens and flow straighteners

Test section

Axial fan

## Solution

A system curve has to be generated and combined with the pressure-discharge characteristics of the fan. Writing the energy equation from the intake of the wind tunnel to the exit

$$\frac{p_1}{\gamma} + \frac{V_1^2}{2g} + z_1 + h_p = \frac{p_2}{\gamma} + \frac{V_2^2}{2g} + z_2 + h_t + h_L$$

which simplifies to

$$h_p = \frac{V_2^2}{2g} + h_L$$

or in terms of pressure

$$\Delta p_p = \rho \frac{V_2^2}{2} + \Delta p_L$$

The velocity at the exit can be expressed in terms of discharge as

$$V_2 = \frac{Q}{A} = \frac{Q}{\frac{\pi}{4} \times 1^2} = 1.27 Q$$

The pressure loss is given by

$$\Delta p_L = 0.2 \rho \frac{V_i^2}{2} + 0.05 \rho \frac{V_2^2}{2}$$

where $V_i$ is the inlet velocity and related to discharge by

$$V_i = \frac{Q}{\frac{\pi}{4} \times 1.5^2} = 0.556Q$$

Substituting into the equation for pressure across the fan

$$\Delta p_p = 1.2 \times (1.05 \times \frac{1.27^2 Q^2}{2} + 0.2 \times \frac{0.556^2 Q^2}{2})$$
$$= 1.05 Q^2$$

Equating this to the pressure-discharge curve

$$1.05 Q^2 = 1000 \times \left[1 - \left(\frac{Q}{100}\right)^2\right]$$

and solving for discharge

$$Q = 29.5 \text{ m}^3/\text{s}$$

The velocity in the test section is

$$V = 1.27 Q = \underline{\underline{37.5 \text{ m/s}}}$$

# Chapter 15

# Varied Flow in Open Channels

## Problem 15.1

Water flows in a circular concrete pipe (Manning's $n = 0.012$) with a depth that is half of the pipe diameter (0.8 m). If the slope is 0.004, find the flow rate.

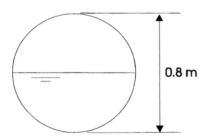

### Solution

The flow rate is obtained from the Chezy equation.

$$Q = \frac{1.0}{n} A R_h^{2/3} S_o^{1/2}$$

The flow area is

$$A = \frac{1}{2} \times \frac{\pi}{4} \times 0.8^2 = 0.251 \text{ m}^2$$

The wetted perimeter is

$$P = \frac{\pi}{2} \times 0.8 = 1.26 \text{ m}$$

The hydraulic radius is

$$R_h = \frac{A}{P} = \frac{0.251}{1.26} = 0.2 \text{ m}$$

The flow rate is

$$Q = \frac{1}{0.012} \times 0.251 \times 0.2^{2/3} \times 0.004^{1/2}$$
$$= \underline{\underline{0.45 \text{ m}^3/\text{s}}}$$

# Problem 15.2

A troweled concrete ($n = 0.012$) open channel has a cross-section as shown. The discharge is 400 cfs. The drop of the channel is 10 ft in each horizontal mile (5280 ft). Find the depth of the flow, $h$.

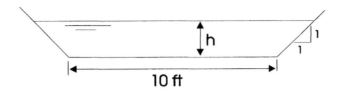

## Solution

The flow rate in traditional units is

$$Q = \frac{1.49}{n} A R_h^{2/3} S_o^{1/2}$$

The slope is

$$S_o = \frac{10}{5280} = 0.00189$$

Thus

$$A R_h^{2/3} = \frac{nQ}{1.49 S_o^{1/2}} = \frac{0.012 \times 400}{1.49 \times 0.00189^{1/2}} = 74.1 \text{ ft}^{8/3}$$

The flow area in terms of depth is

$$A = (10 + h)h$$

The wetted perimeter is

$$P = 10 + 2\sqrt{2}h$$

so the hydraulic radius is

$$R_h = \frac{A}{P} = \frac{(10+h)h}{10 + 2\sqrt{2}h}$$

Thus

$$A R_h^{2/3} = \frac{(10+h)^{5/3} h^{5/3}}{(10 + 2\sqrt{2}h)^{2/3}} = 74.1$$

or

$$\frac{(10+h)h}{(10 + 2\sqrt{2}h)^{2/5}} = 13.24$$

Solving this equation by iteration gives

$$h = \underline{3.26 \text{ ft}}$$

## Problem 15.3

Find the flow rate in the channel and overbank area that is shown in the following figure. The slope of the channel is 0.001, and the depth in the overbank area is 2 m. The Manning's n is 0.04 in the overbank area and 0.03 in the main channel. All the channel sides have a 1:1 slope.

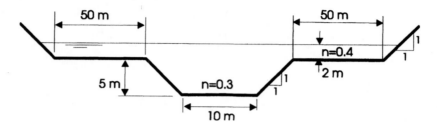

### Solution

The discharge is given by the Chezy equation.

$$Q = \frac{1.0}{n} A R_h^{2/3} S_o^{1/2}$$

For the overbank area

$$A = 2 \times (50 + h_o/2) \times h_o$$
$$= (100 + h_o)h_o$$

where $h_o$ is the depth in the overbank area. So when $h_o = 2$ m, the area is $A = 204$ m$^2$.

The wetted perimeter is

$$P = 2 \times (50 + \sqrt{2}h_o)$$
$$= 100 + 2\sqrt{2}h_o$$

So when $h_o = 2$ m, the wetted perimeter is $P = 105.6$ m.

The hydraulic radius for the overbank area is

$$R_h = \frac{A}{P} = \frac{204}{105.7} = 1.93 \text{ m}$$

For the main channel

$$A = 10h_c + 2 \times 5(h_c - 5) + 2 \times 5 \times 5/2$$
$$= 20h_c - 25$$

With the main channel depth being 7 m, the flow area is 115 m². The wetted perimeter is
$$P = 10 + 2 \times \sqrt{2} \times 5$$
$$= 24.1 \text{ m}$$
so the hydraulic radius of the main channel is
$$R_h = \frac{A}{P} = \frac{115}{24.1} = 4.77 \text{ m}$$
The flow rate is the sum in each area.
$$Q = \frac{1}{0.04} \times 204 \times 1.93^{2/3} \times 0.001^{1/2} + \frac{1}{0.03} \times 115 \times 4.77^{2/3} \times 0.001^{1/2}$$
$$= 250.0 + 343.5$$
$$= \underline{\underline{593.5 \text{ m}^3/\text{s}}}$$

## Problem 15.4

Water with a depth of 15 cm and a speed of 6 m/s flows through a rectangular channel. Determine if the flow is critical, subcritical, or supercritical. If appropriate, determine the alternative depth.

### Solution

The nature of the flow is determined by the Froude number.
$$Fr = \frac{V}{\sqrt{gD}}$$
$$= \frac{6 \text{ m/s}}{\sqrt{\left(9.8 \text{ m/s}^2\right)(0.15 \text{ m})}}$$
$$= 4.95$$

Since $Fr > 1$, the flow is supercritical. To find the alternative depth, note that the specific energy of subcritical and supercritical flow are the same.
$$\left(y + \frac{V^2}{2g}\right)_1 = \left(y + \frac{V^2}{2g}\right)_2 \quad (1)$$
$$= \left(0.15 + \frac{6^2}{2 \times 9.8}\right)$$
$$= 1.99 \text{ m}$$

where subscripts 1 and 2 denote sub- and supercritical, respectively. To solve Eq. (1) for subcritical depth ($y_1$), speed is needed. The continuity principle gives
$$(VA)_1 = (VA)_2$$
$$(Vyw)_1 = (Vyw)_2$$

so

$$V_1 = V_2 \frac{y_2}{y_1} \qquad (2)$$
$$= 6\frac{0.15}{y_1}$$
$$= \frac{0.9}{y_1}$$

Combining Eqs. (1) and (2) gives

$$\left(y + \frac{V^2}{2g}\right)_1 = 1.99 \text{ m}$$
$$y_1 + \frac{0.9^2/y_1^2}{2 \times 9.8} = 1.99$$

or

$$y_1^3 - 1.99y_1^2 + 0.04133 = 0$$

We solved for the roots of this cubic equation using a computer program (MathCad). The solution has three roots: $y_1 = (-0.139, 0.15, 1.979 \text{ m})$. Thus the alternate depth is

$$y_1 = \underline{1.979 \text{ m}}$$

## Problem 15.5

Water flows at a uniform rate of 400 cfs through a rectangular channel that has a slope of 0.007 and a width of 25 ft. The channel sides are concrete with a roughness factor of $n = 0.015$. Determine depth of flow, and whether the flow is critical, subcritical, or supercritical.

### Solution

The nature of the flow is determined by the Froude number.

$$Fr = \frac{V}{\sqrt{gy}}$$

To find the depth $y$, we can use Manning's equation.

$$Q = \frac{1.49}{n} A R_h^{3/2} S_o^{1/2}$$

$$400 = \frac{1.49}{0.015}(25 \times y)\left(\frac{25y}{25+2y}\right)^{3/2}(0.007)^{1/2}$$

To solve for $y$ in Manning's equation, we used a computer program (MathCad) to find a root for an equation of the form $f(x) = 0$. The result is

$$\underline{\underline{y = 1.38 \text{ ft}}}$$

Discharge is

$$Q = VA = Vyw$$
$$(400 \text{ ft}^3/\text{s}) = V(1.38 \times 25 \text{ ft}^2)$$

so $V = 11.59$ ft/s.

The Froude number is

$$Fr = \frac{V}{\sqrt{gy}}$$
$$= \frac{11.59 \text{ ft/s}}{\sqrt{(32.2 \text{ ft/s}^2)(1.38 \text{ ft})}}$$
$$= 1.74$$

Thus

$$\underline{\text{flow is supercritical}}$$

## Problem 15.6

Water flows in a rectangular channel that ends in a free outfall. The channel has a slope of 0.005, a width of 20 ft, and a depth at the brink of 2 ft. Find the discharge in the channel.

## Solution

A sketch of the situation is

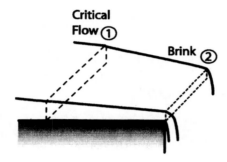

At the brink, the depth is 71% of critical depth.

$$y_1 = \frac{y_2}{0.071}$$
$$= \frac{2 \text{ ft}}{0.71}$$
$$= 2.82 \text{ ft}$$

At section 1, the flow is critical, so the Froude number is 1.0.

$$1.0 = \frac{V}{\sqrt{gy}} \qquad (1)$$

From continuity

$$Vy = q \qquad (2)$$

Combining Eqs. (1) and (2) gives

$$q = \sqrt{gy^3}$$
$$= \sqrt{\left(32.2 \text{ ft/s}^2\right)\left(2.82^3 \text{ ft}^3\right)}$$
$$= 26.9 \text{ ft}^2/\text{s}$$

Thus

$$Q = qw$$
$$= (26.9 \text{ ft}^2/\text{s})(20 \text{ ft})$$
$$= \underline{538 \text{ cfs}}$$

---

## Problem 15.7

Water flows with an upstream velocity of 6 ft/s and a depth of 12 ft in a rectangular open channel. The water passes over a gradual 18-in. upstep. Determine the depth of the water and the change in surface level downstream of the upstep.

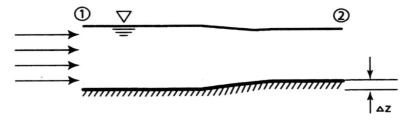

---

### Solution

Assuming no energy losses, the specific energy is constant across the upstep.

$$y_1 + \frac{V_1^2}{2g} = y_2 + \frac{V_2^2}{2g} + \Delta z \tag{1}$$

The continuity principle is

$$y_1 V_1 = y_2 V_2$$

So

$$V_2 = V_1 \frac{y_1}{y_2} \tag{2}$$

Combining Eqs. (1) and (2)

$$y_1 + \frac{V_1^2}{2g} = y_2 + \frac{V_1^2}{2g}\left(\frac{y_1}{y_2}\right)^2 + \Delta z$$

$$12 \text{ ft} + \left(\frac{6^2 \text{ ft}^2/\text{s}^2}{2 \times 32.2 \text{ ft/s}^2}\right) = y_2 + \left(\frac{6^2 \text{ ft}^2/\text{s}^2}{2 \times 32.2 \text{ ft/s}^2}\right)\left(\frac{12 \text{ ft}}{y_2}\right)^2 + 18/12 \text{ ft}$$

So

$$11.06 = y_2 + \frac{80.50}{y_2^2}$$

or

$$y_2^3 - 11.06 y_2^2 + 80.5 = 0$$

Solving this cubic equation using a computer program (MathCad) gives three roots $y_2 = (-2.442, 3.2, 10.30)$. The negative root is not possible, and the small root (supercritial flow) is unlikely. Thus, the depth of water at section 2 is

$$y_2 = \underline{10.3 \text{ ft}}$$

The elevation of the water surface at section 2 is the sum of the depth of the water and the height of the upstep.

$$\begin{aligned} z_2 &= y_2 + \Delta z \\ &= 10.3 \text{ ft} + 1.5 \text{ ft} \\ &= \underline{11.8 \text{ ft}} \end{aligned}$$